SCHOLAR Study Guide

CfE Advanced Higher

Unit 1: Inorganic and Phys [...] hemistry

Authored by:

Diane Oldershaw (Menzieshill High School)

Reviewed by:

Helen McGeer (Firrhill High School)

Nikki Penman (The High School of Glasgow)

Previously authored by:

Peter Johnson

Brian T McKerchar

Arthur A Sandison

Heriot-Watt University

Edinburgh EH14 4AS, United Kingdom.

First published 2016 by Heriot-Watt University.

This edition published in 2016 by Heriot-Watt University SCHOLAR.

Distributed by the SCHOLAR Forum.

SCHOLAR Study Guide Unit 1: CfE Advanced Higher Chemistry

1. CfE Advanced Higher Chemistry Course Code: C713 77

ISBN 978-1-911057-33-8

Printed and bound by CPI Group (UK) Ltd, Croydon, CR0 4YY

Acknowledgements

Thanks are due to the members of Heriot-Watt University's SCHOLAR team who planned and created these materials, and to the many colleagues who reviewed the content.

We would like to acknowledge the assistance of the education authorities, colleges, teachers and students who contributed to the SCHOLAR programme and who evaluated these materials.

Grateful acknowledgement is made for permission to use the following material in the SCHOLAR programme:

The Scottish Qualifications Authority for permission to use Past Papers assessments.

The Scottish Government for financial support.

The content of this Study Guide is aligned to the Scottish Qualifications Authority (SQA) curriculum.

Contents

Topic 1

Electromagnetic radiation and atomic spectra

Contents

Prerequisite knowledge

Before you begin this topic, you should know:

- *electronic structure of atoms (National 4/5+ Higher Chemistry Unit 1 Chemical Changes and Structure).*

Learning objectives

By the end of this topic, you should know:

- *there is a spectrum of electromagnetic radiation;*
- *electromagnetic radiation can be described in the terms of waves;*
- *electromagnetic radiation can be characterised in terms of wavelength or frequency;*
- *the relationship between wavelength and frequency is given by* $c = \lambda \times f$;
- *absorption or emission of electromagnetic radiation causes it to behave more like a stream of particles called photons;*
- *the energy lost or gained by electrons associated with a single photon is given by* $E = hf$;
- *it is more convenient for chemists to express the energy for one mole of photons as* $E = Lhf$ *or*

$$E = \frac{Lhc}{\lambda}$$

- *atomic emission spectra are made up of lines at discrete frequencies;*
- *photons of light energy are emitted by atoms when electrons move from a higher energy level to a lower one;*
- *photons of light energy are absorbed by atoms when electrons move from a lower energy level to a higher one;*
- *each element produces a unique pattern of frequencies of radiation in its emission and absorption spectra;*
- *atomic emission spectroscopy and atomic absorption spectroscopy are used to identify and quantify the elements present in a sample;*
- *write electronic configurations using spectroscopic notation.*

1.1 Electromagnetic radiation

Electromagnetic radiation is a form of energy. Light, x-rays, radio signals and microwaves are all forms of electromagnetic radiation. Visible light is only a small part of the range of the **electromagnetic spectrum**.

Figure 1.1: The electromagnetic spectrum

....

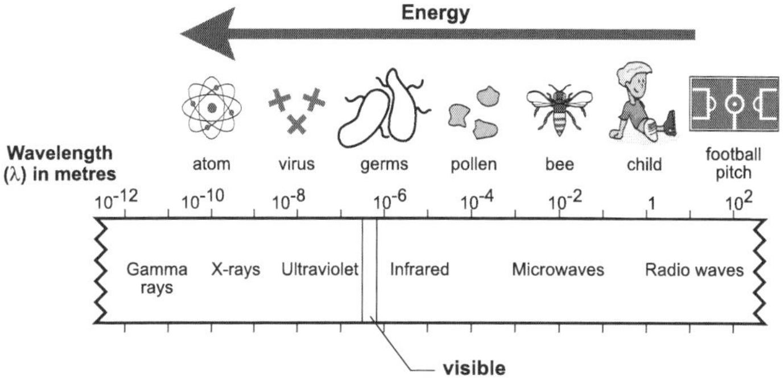

The figure above shows that the electromagnetic spectrum has a variety of ways in which it can be described. At the highest energy, the waves are so tightly packed that they are less than an atom's width apart, whilst at the low energy end the waves are a football pitch or greater apart. In the wave model description of electromagnetic radiation, the waves can be specified by their wavelength and frequency. All electromagnetic radiation travels at the same **velocity** . This constant is the speed of light, symbol 'c', and, in a vacuum, it is approximately equal to

$$c = 3 \times 10^{8} \text{ m s}^{-1}$$

In the waveform diagram below, two waves are travelling with velocity c. In one second, they travel the same distance.

Figure 1.2: Waveform diagram

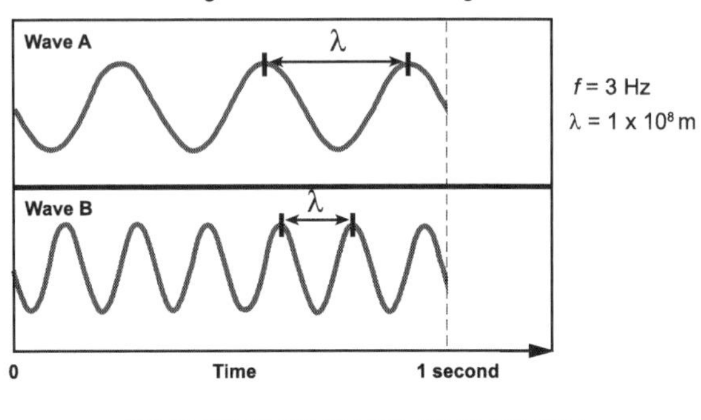

....

Formulae are found on page four of the SQA CfE Higher and Advanced Higher Chemistry data booklet.

Wavelength has the symbol λ (lambda). It is the distance between adjacent crests (or troughs) and is usually measured in metres or nanometres (1 nm = 10^{-9} m). In the waveform diagram above, wave A has a wavelength twice the value of wave B.

Frequency has the symbol *f*. It is the number of wavelengths that pass a fixed point in one unit of time, usually one second. Frequency is measured as the reciprocal of time (s^{-1}), more commonly called 'hertz' (Hz). In the waveform diagram above, wave A has half the frequency value of wave B.

Frequency and wavelength are very simply related. Multiplying one by the other results in a constant value called '*c*', the speed of light.

c = wavelength × *frequency*

$$c = \lambda \times f$$

The wavelength of wave B (in the waveform diagram above) can be checked using this equation.

$$c = \text{ speed of light} = 3 \times 10^8 \text{ m s}^{-1}$$
$$f = \text{frequency} = 6 \text{ Hz (or s}^{-1})$$
$$c = \lambda \times f$$
$$\lambda = \frac{c}{f}$$
$$\lambda = \frac{3 \times 10^8 \text{ m s}^{-1}}{6 \text{ s}^{-1}}$$
$$\lambda = 5 \times 10^7 \text{ m}$$

Notice that this is half the value given for wave A (in the waveform diagram above).

Try **Question 1**, basing your method on the last example shown. The full working is available in the answer at the end of this booklet and can be used if you have difficulty. You are strongly advised to try these questions on paper. Seeing the solution is not the same as solving the problem!

Obtaining wavelength from frequency

Q1: A typical microwave oven operates at a frequency of 2.45 x 10^9 Hz.
Calculate the wavelength of this radiation. Give your answer in centimetres.

....

Q2: A beam of light from a sodium street lamp is found to have a frequency of 5.09 $\times$ 10^{14} Hz.
Calculate the wavelength of this light to the nearest nanometre.

....

Wavelength from frequency

Go online

Q3: Electromagnetic radiation is found to have a frequency of 8 $\times$ 10^{12} Hz.
Calculate the wavelength of this radiation. Give your answer *to three significant figures.*

....

Q4: Electromagnetic radiation is found to have a frequency of 7 $\times$ 10^{14} Hz.
Calculate the wavelength of this radiation. Give your answer to *three significant figures.*

....

Q5: Electromagnetic radiation is found to have a frequency of 8 $\times$ 10^9 Hz.
Calculate the wavelength of this radiation. Give your answer to *three significant figures.*

....

In other situations, the wavelength may be given and the value of frequency can be calculated by rearranging the equation to

$$f = \frac{c}{\lambda}$$

Obtaining frequency from wavelength

Q6: Data book information records the flame colour of potassium as lilac and with a wavelength of 405 nm.
Calculate the frequency of this radiation.

....

Q7: Use the data book to find the wavelength of light emitted by a sample of copper in a flame and thus calculate its frequency. The frequency in hertz is:

a) 9.23×10^{17}
b) 1.08×10^{12}
c) 9.23×10^{14}
d) 1.08×10^{16}

....

Go online

Frequency from wavelength

Q8: Electromagnetic radiation is found to have a wavelength of 1900 nm.
Calculate the frequency of the radiation. Give your answer to *three significant figures.*

....

Q9: Electromagnetic radiation is found to have a wavelength of 2300 nm.
Calculate the frequency of the radiation. Give your answer to *three significant figures.*

....

Q10: Electromagnetic radiation is found to have a wavelength of 1500 nm.
Calculate the frequency of the radiation. Give your answer to *three significant figures.*

....

Go online

Electromagnetic radiation table

This is a summary table showing the relationship between the symbols for the various radiation characteristics and their units and descriptions.

Q11: Complete the table by putting the boxes into the correct column.

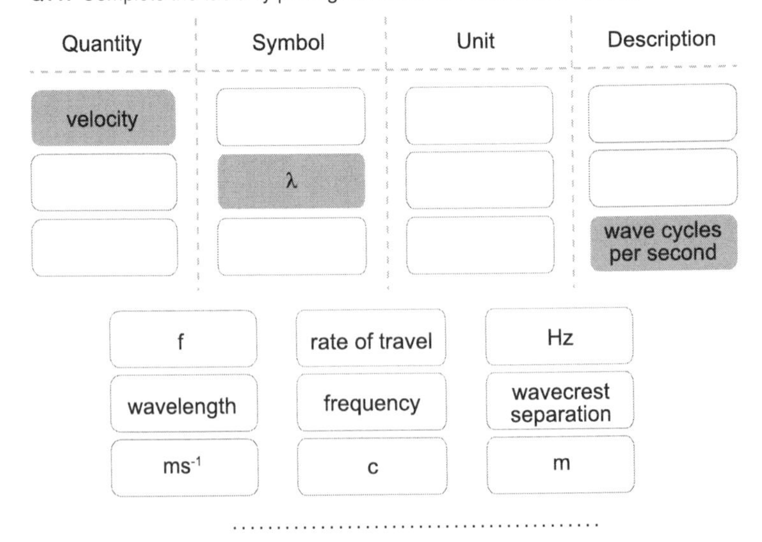

....

1.2 Spectroscopy

Visible light is only a small part of the electromagnetic spectrum. A wider range would stretch from gamma rays to radio and TV waves (Figure 1.3).

Figure 1.3: Electromagnetic spectrum

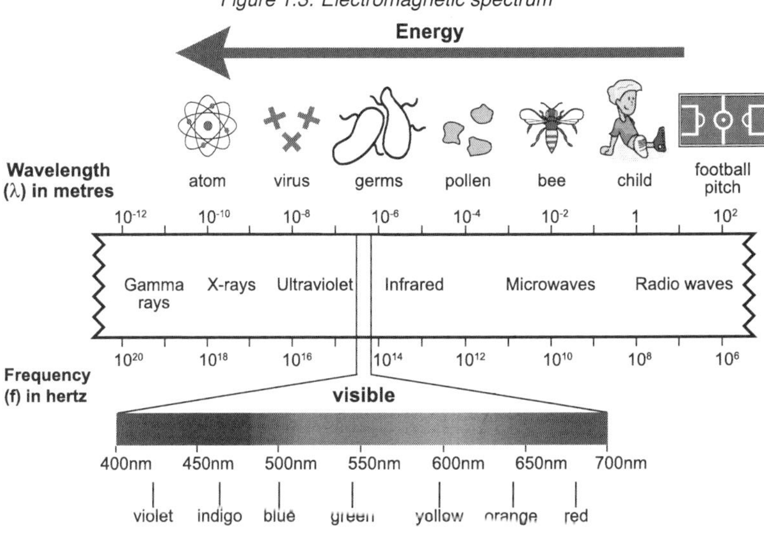

When a beam of white light is passed through a prism or from a diffraction grating onto a screen, a continuous spectrum is seen Figure 1.4 (a). The same effect can be seen in a rainbow.

However, if the light source is supplied by sodium chloride being heated in a Bunsen burner flame, the spectrum turns out not to be a continuous spectrum, but a series of lines of different wavelengths and thus of different colours.

Spectra that show energy being given out by an atom or ion are called atomic emission spectra as shown in Figure 1.4 (b). The pattern of lines in such a spectrum is characteristic of each element and, like a fingerprint, can be used to identify the element.

Figure 1.4: Types of spectrum

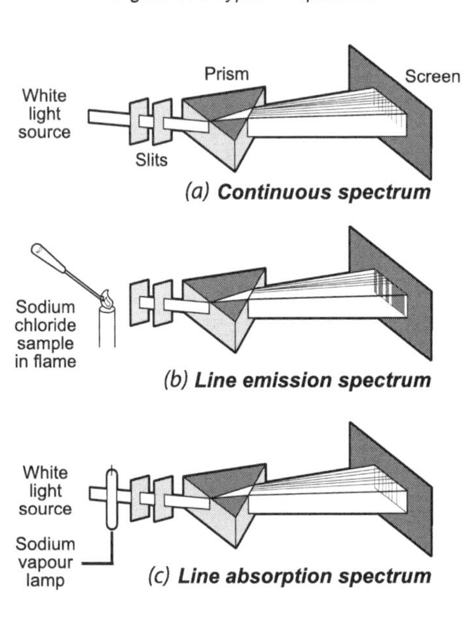

....

If a beam of continuous radiation like white light is directed through a gaseous sample of an element, the radiation that emerges has certain wavelengths missing. This shows up as dark lines on a continuous spectrum and is called an atomic absorption spectrum, see Figure 1.4 (c).

This also provides a pattern that can often be used in identification. In both techniques some lines normally occur in the visible region (400-700 nm) but some applications use the ultraviolet region (200-400 nm). Both emission and absorption spectroscopy can be used to determine whether a certain species is present in a sample and how much of it is present, since the intensity of transmitted or absorbed radiation can be measured.

In atomic absorption spectra electromagnetic radiation is directed at an atomised sample. The electrons are promoted to higher energy levels as the radiation is absorbed. The absorption spectrum is a measure of the sample's transmission of light at various wavelengths.

In atomic emission spectra high temperatures are used to excite electrons within the atoms. As they fall back down to their original energy level photons are emitted. The emission spectrum is a measure of the light emitted by the sample at different wavelengths.

Absorption spectrometer

An online and printable electromagnetic spectrum is available from the Royal Society of Chemistry.

View the videos on spectroscopy on the Royal Society of Chemistry's website.
http://www.rsc.org/learn-chemistry/resource/res00001041/spectroscopy-videos

. . . .

1.3 Using spectra to identify samples

Spectra can be used to give information about how much of a species is present in a sample. For example the concentration of lead in drinking water or a foodstuff can be found. First a calibration graph is prepared from known concentrations of lead solutions. The radiation absorbed by these samples is plotted against concentration and when the unknown sample is analysed the concentration of lead can be found from the graph.

Figure 1.5: Analysing an unknown concentration by spectroscopy

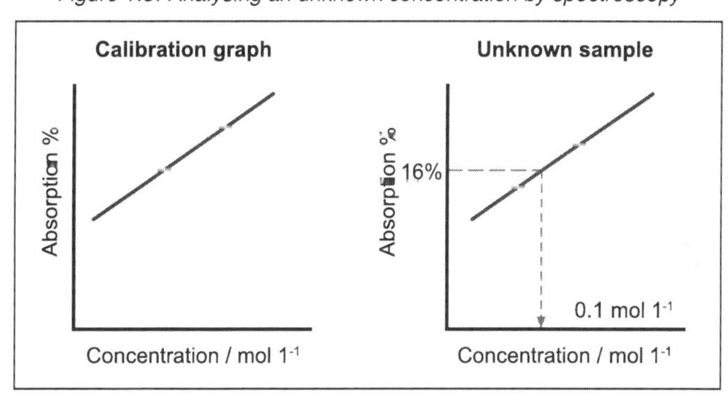

. . . .

The figure above shows the build up of a calibration graph on the left as the radiation absorbed is measured at different concentrations. After the graph is complete, the unknown sample is measured (in this case at 16% absorption) and reading off from the graph shows a concentration in the sample of 0.1 mol l^{-1}.

Using spectra to identify samples

In this activity you can use the online database or the printed version (Figure 1.6) to answer the questions about these two spectral problems.

Figure 1.6: Spectral database

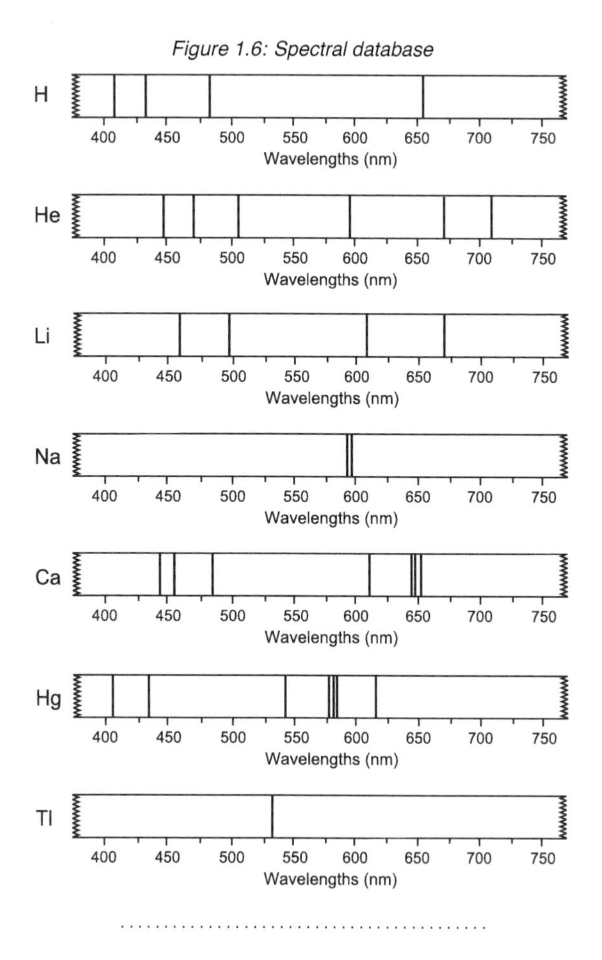

....

Part a) This spectrum was obtained from the atmosphere around the Sun.

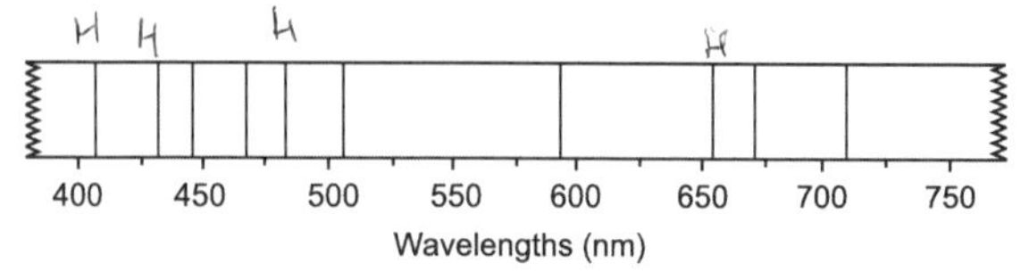

Q12: Write the name of the element which is responsible for the line at 655 nm.

. . . .

Q13: Write the name of the other element present.

. . . .

Part b) These were obtained from equal sized soil samples that have been treated and the spectra measured.

1. ***Sample A.*** From productive farmland
2. ***Sample B.*** From the site of an old factory site where insecticides and pesticides were produced.

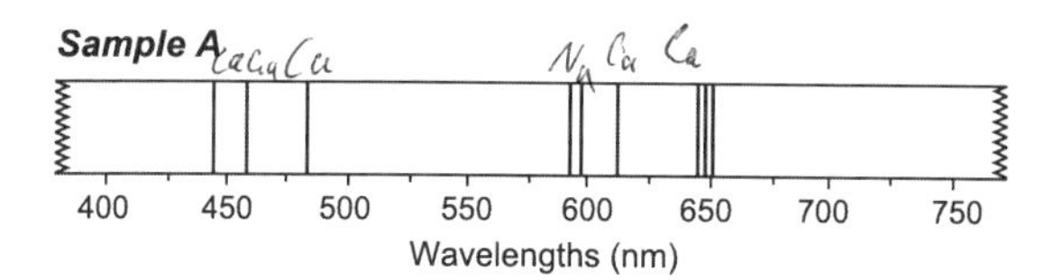

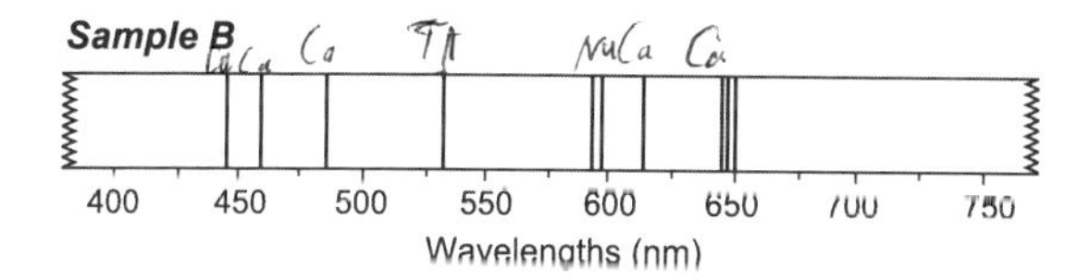

The boxes in this grid contain elements whose spectra appear in Figure 1.6.

A	B	C
Li	Hg	He
D	**E**	**F**
Tl	Na	Ca

Q14: Which element is responsible for the triplet of lines around 650 nm?

. . . .

Q15: Which other metal is present in ***both*** samples?

. . . .

Q16: Use the database (Figure 1.6) to work out which element is present in sample (B) but not in (A).

. . . .

Q17: Mercury is a metal whose salts are well known poisons and thallium salts are used in some countries as a rat poison. Is there evidence of mercury in sample (A)?

....

Q18: Is there evidence of mercury in sample (B)?

....

Q19: Write the ***name*** of the element which you think the spectra prove to be the cause of the pollution.

....

....

1.4 Energy calculations

The full emission spectrum for the hydrogen atom consists of a number of series of lines, named after the scientists who first investigated the spectra. Only the Balmer series lies in the visible region.

Figure 1.7: Hydrogen spectrum

The Hydrogen Spectrum
(showing the different series of lines)

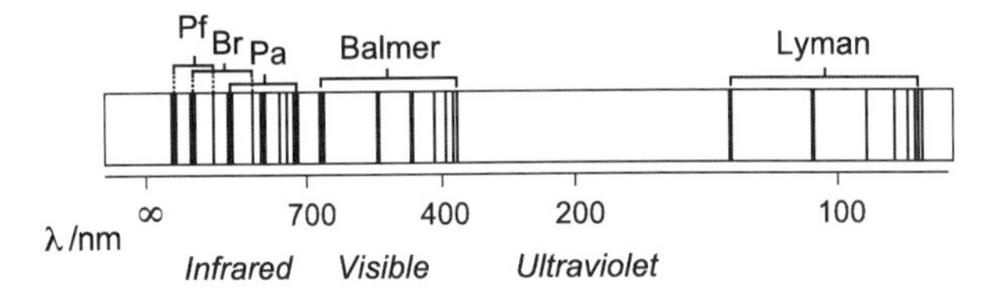

(Pf = Pfund Br = Brackett Pa = Paschen Br overlaps Pf and Pa)

....

The spectral lines of radiation emitted by the hydrogen atom in the spectrum (Figure 1.7) show emission at only certain frequencies. Since electromagnetic radiation carries energy related to the frequency, only certain precise energy values are being involved.

Max Planck developed the theory that under certain circumstances electromagnetic radiation may be regarded as a stream of particles. These particles are called photons.

The energy carried by a photon is related to its frequency by the equation:

$$E = hf$$

where h = **Planck's constant** = 6.63×10^{-34} J s

So when dealing with one mole of photons the energy involved would be:

$$E = Lhf$$

where L = **Avogadro's constant** = 6.02 x 10^{23} mol^{-1}

$$\text{but } c = \lambda \times f$$
$$\text{so } f = \frac{c}{\lambda}$$
$$\text{and } E = \frac{Lhc}{\lambda}$$

Each line in a spectrum has a precise frequency that corresponds to a fixed value of energy. Calculations using these formulae are common.

Example : Obtaining energy values

The red line in the hydrogen spectrum has a wavelength of 656 nm. Calculate:

1. The energy value of one photon of light at this wavelength.
2. The energy value in kJ mol^{-1} for one mole of photons at this wavelength.

$$c = \lambda \times f$$
$$\text{therefore } f = \frac{c}{\lambda}$$
$$\text{and since } E = hf \text{ for one photon}$$
$$\text{then } E = \frac{hc}{\lambda}$$
$$E = \frac{6.63 \times 10^{-34} \text{ J s} \times 3 \times 10^{8} \text{ m s}^{-1}}{656 \times 10^{-9} \text{ m}}$$
$$E = 3.03 \times 10^{-19} \text{ J (this is for one photon)}$$

So for one mole of photons

$$E = 3.03 \times 10^{-19} \text{ J} \times 6.02 \times 10^{23} \text{ mol}^{-1}$$
$$E = 18.24 \times 10^{4} \text{ J mol}^{-1}$$
$$E = 1.82 \times 10^{5} \text{ J mol}^{-1}$$
$$E = 182 \text{ kJ}$$

. . . .

Q20: Chlorinated hydrocarbon molecules contain **C-Cl** bonds. The energy required to break these is 338 kJ mol^{-1}.

1. Calculate the wavelength of light required to break one mole of these bonds.
2. By reference to the electromagnetic spectrum, suggest why these molecules are unstable in the upper atmosphere.

....

Go online

Energy from wavelength

Q21: The line in an emission spectrum has a wavelength of 1600 nm.
Calculate the energy value for one mole of photons at this wavelength. Give your answer to one decimal place.

....

Q22: The line in an emission spectrum has a wavelength of 1300 nm.
Calculate the energy value for one mole of photons at this wavelength. Give your answer to one decimal place.

....

Q23: The line in an emission spectrum has a wavelength of 1400 nm.
Calculate the energy value for one mole of photons at this wavelength. Give your answer to one decimal place.

....

Q24: The line in an emission spectrum has a wavelength of 1900 nm.
Calculate the energy value in kJ mol^{-1} for one mole of photons at this wavelength. Give your answer to one decimal place.

....

....

1.5 Summary

Summary

You should now be able to state that:

- there is a spectrum of electromagnetic radiation;
- electromagnetic radiation can be described in the terms of waves;
- electromagnetic radiation can be characterised in terms of wavelength or frequency;
- the relationship between wavelength and frequency is given by $c = \lambda \times f$;
- absorption or emission of electromagnetic radiation causes it to behave more like a stream of particles called photons;
- the energy lost or gained by electrons associated with a single photon is given by $E = hf$;
- it is more convenient for chemists to express the energy for one mole of photons as $E = Lhf$ or
$$E = \frac{Lhc}{\lambda}$$
- atomic emission spectra are made up of lines at discrete frequencies;
- photons of light energy are emitted by atoms when electrons move from a higher energy level to a lower one;
- photons of light energy are absorbed by atoms when electrons move from a lower energy level to a higher one;
- each element produces a unique pattern of frequencies of radiation in its emission and absorption spectra;
- atomic emission spectroscopy and atomic absorption spectroscopy are used to identify and quantify the elements present in a sample.

1.6 Resources

- Royal Society of Chemistry (http://www.rsc.org)
- SSERC (http://bit.ly/29QE71h) - activity with filter paper soaked in brine to observe sodium spectrum.
- 800mainstreet.com (http://bit.ly/1IWIDAN) - a useful resource on spectroscopy and the identification of elements from emission spectra.
- Nuffield Foundation (http://bit.ly/2a79bNI) - spectra formed by gratings.
- Experiment 10: Colorimetric determination of manganese in steel (http://bit.ly/29Vgnv0) (from 'Chemistry: A Practical Guide Support Materials').

1.7 End of topic test

End of Topic 1 test

Go online

The end of topic test for *Electromagnetic radiation and atomic spectra*

Q25: Which of these has the highest frequency?

a) γ - radiation
b) α - radiation
c) Visible light
d) Radio waves

....

Q26: Electromagnetic radiation may be regarded as a stream of:

a) x-rays.
b) α-particles.
c) electrons.
d) photons.

....

Q27: Which of these happens as the frequency of an orange laser light is decreased?

a) Wavelength goes down.
b) Colour moves towards red.
c) Velocity goes down
d) Energy increases.

....

Q28: Compared to infrared radiation, ultraviolet radiation has:

a) lower frequency.
b) lower velocity.
c) higher energy.
d) higher wavelength.

....

Q29: Which of the following is used to represent velocity?

a) f
b) λ
c) c
d) E
e) h

....

Q30: Which of the following is used to represent a constant? Choose two.

a) f
b) λ
c) c
d) E
e) h

....

Q31: Which of the following is measured in Hertz?

a) f
b) λ
c) c
d) E
e) h

....

Look at this emission spectrum for hydrogen.

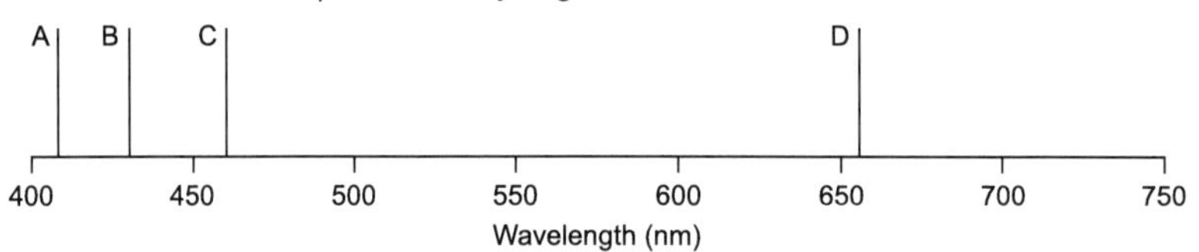

Q32: Which of the lines has the lowest energy (A, B, C, or D)?

a) A
b) B
c) C
d) D

....

Q33: Calculate the wavelength, to the nearest nanometre, of the line with a frequency of 7.35×10^{14} Hz.

....

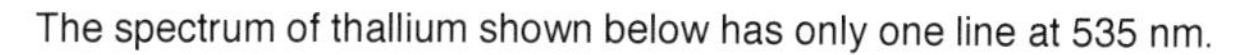

The spectrum of thallium shown below has only one line at 535 nm.

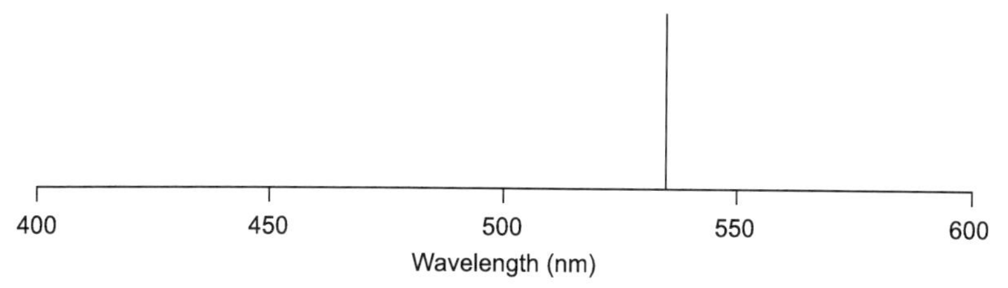

Q34: What colour would thallium salts show up as in a Bunsen flame?

....

Q35: Calculate the energy in $kJmol^{-1}$ of the emission line that occurs at 535 nm. Give your answer to one decimal place.

....

....

Topic 2

Atomic orbitals, electronic configurations and the Periodic Table

Contents

Prerequisite knowledge

Before you begin this topic, you should understand:

- *the structure of the atom in terms of the subatomic particles (National 4 and National 5 - Unit 1 Chemical Changes and Structure);*
- *the electromagnetic spectrum, emission and absorption spectroscopy (Unit 1, Topic 1, Advanced Higher Chemistry);*
- *how to solve problems and calculations associated with the electromagnetic spectrum, emission and absorption spectroscopy (Unit 1 Topic 1 Advanced Higher Chemistry);*
- *ionisation energy (Unit 1 Chemical Structures and Changes, CfE Higher Chemistry).*

Learning objectives

By the end of this topic, you should be able to:

- *relate spectral evidence to electron movements and ionisation energy;*
- *describe the four quantum numbers and relate these to atomic orbitals, their shape and relative energies;*
- *describe the four s, p and d atomic orbitals, their shapes and relative energies;*
- *relate the ionisation energies of elements to their electronic configuration, and therefore to their position in the Periodic Table;*
- *describe the electronic configuration of atoms 1-20 in spectroscopic notation;*
- *write electron configurations in spectroscopic notation.*

2.1 Spectra, quanta and ionisation

Light can interact with atoms and provide valuable information about the quantity and type of atom present. Electromagnetic radiation has provided important clues about the actual structure of the atom and the organisational relationship between the elements in the Periodic Table.

The spectral lines in an atomic emission or absorption spectrum occur at precise frequencies. Since the frequency is related to energy it is obvious that only certain precise energy values are involved (see Figure 2.1).

Figure 2.1: Hydrogen atom spectrum

The Hydrogen Spectrum
(showing the different series of lines)

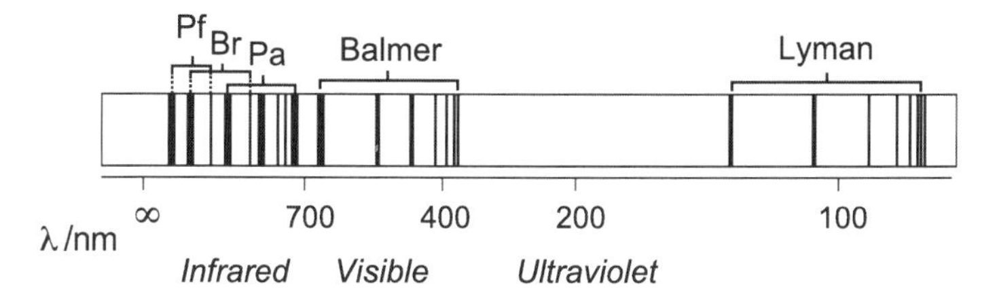

(Pf = Pfund Br = Brackett Pa = Paschen Br overlaps Pf and Pa)

. . . .

Atomic spectra are caused by electrons moving between different energy levels. These are fixed for any one atom. We say that the energy of electrons in atoms is quantised. Quantum theory states that matter can only emit or absorb energy in small fixed amounts. When an electron in an atom absorbs a photon of energy, it moves from a lower energy level to a higher energy level. When the electron drops back down, energy is emitted (see Figure 2.2).

Figure 2.2: Emission of energy

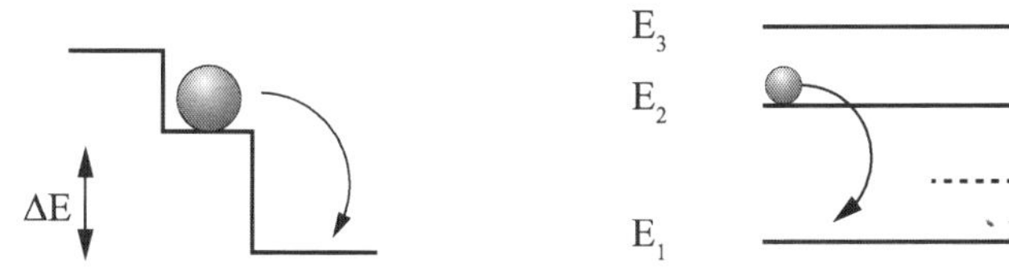

The ball has certain stable levels. Energy is emitted if it falls down.

The electron has discrete energies. Energy is emitted if it falls down.

. . . .

The energy of the photon emitted is

$$\Delta E = E_2 - E_1 = hf$$

The frequency of the line in the emission spectrum represents the difference in energy between the levels. We call these energy levels shells or sub-shells and in Figure 2.3 the letter n = 1, 2, 3, etc. defines what is known as the principal quantum number.

Figure 2.3: The Balmer Series

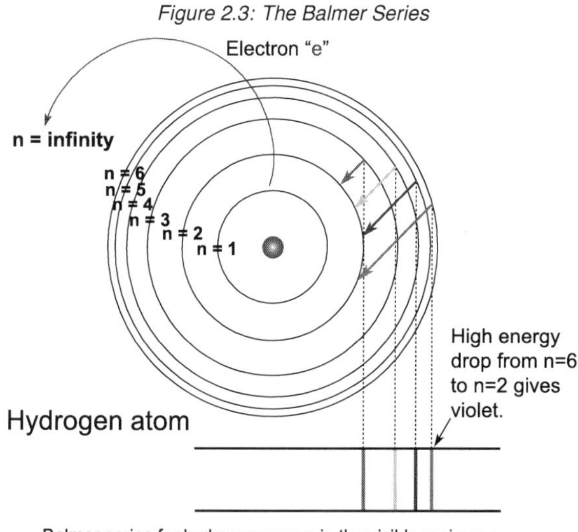

. . . .

Figure 2.3 (the Balmer series for hydrogen) shows the spectrum produced when electrons, having been excited into higher energy levels, drop back to the n = 2 level and emit radiation in the visible region. Electrons dropping to n = 1 level would emit in the ultraviolet (Lyman series). Notice also that the levels get closer together as n increases.

Level number n = 1 is the lowest energy level and an electron in a hydrogen atom would occupy this level in the **ground state** under normal conditions.

In Figure 2.3, an electron labelled as electron '**e**' is shown **escaping** from level n = 1 to infinity. This corresponds to the electron breaking away from the atom completely and represents the **ionisation energy** of that element.

It can be represented in general terms as

$$X_{(g)} \rightarrow X^{+}_{(g)} + e^{-}$$

In the hydrogen atom, the highest energy line in the Lyman series where the lines converge (see Figure 2.1) occurs at a wavelength of 91.2 nm. The ionisation energy can be calculated from this wavelength.

Calculating ionisation energy

Q1: Calculate the ionisation energy in kJ mol^{-1} for the hydrogen atom from the spectral information that the Lyman series converges at 91.2 nm.

Try solving this problem for yourself using the equations from Unit 1, Topic 1.

....

2.2 Quantum numbers

The lines in the spectrum of hydrogen are adequately explained by the picture of the atom in the last section. Emission spectra of elements with more than one electron provide evidence of sub-levels within each principal energy level above the first. Quantum theory now defines the allowed energy levels of electrons by four quantum numbers. No two electrons in an atom can have the same four quantum numbers. These quantum numbers can be thought of as 'addresses' for electrons.

For example: Mr Smith lives at Flat 2, number 8, Queen Street, Perth.

His Quantum address	Perth	Queen Street	Number 8	Flat 2
This defines his position	Town	Street	Number	House

If the same framework is considered for Quantum numbers and electrons.

For an electron in an atom	Principal Quantum number	Second Quantum number	Third Quantum number	Fourth Quantum number
This defines the position	Shell	Sub-shell	Direction	Spin

For an electron in an atom each quantum number requires a bit more explanation.

Principal Quantum number, symbol ***n***, determines the main energy level. It can have values $n = 1, 2, 3, 4$, etc. The numbers determine the size and energy of the shell.

Second Quantum number, symbol ℓ, determines the shape of the sub-shell and is labelled as *s, p, d, f.* This can have values from zero to (n -1). The second quantum number is also known as the angular momentum quantum number.

So if n has a value = 4, then ℓ can take these values:

- value 0 labelled as '*s*' subshell;
- value 1 labelled as '*p*' subshell;
- value 2 labelled as '*d*' subshell;
- value 3 labelled as '*f*' subshell.

So for $n = 4$ there are 4 possible subshells. The use of letters (*s*, *p*, *d*, *f*), instead of numbers, aids identification of the sub-shells.

The relationship between n and ℓ is shown in Table 2.1. A useful memory aid is that there are as many subshells as the value of n.

*Table 2.1: Relating **n** and ℓ*

Principal Quantum number value n	Second Quantum number value: ℓ	Sub-shell name
1	0	1*s*
2	0	2*s*
	1	2*p*
3	0	3*s*
	1	3*p*
	2	3*d*

....

The third and fourth quantum numbers will be explained in the next section.

2.3 Atomic orbitals

Before determining where an electron is within a sub-shell and considering the third quantum number, it must be understood that electrons display the properties of both particles and waves. If treated as particles, **Heisenberg's uncertainty principle** states that it is impossible to state precisely the position and the momentum of an electron at the same instant. If treated as a wave, the movement of an electron round the nucleus can be described mathematically. From solutions to these wave equations it is possible to produce a statistical picture of the probability of finding electrons within a given region. Regions of high probability are called atomic orbitals.

Figure 2.4: Shapes of s orbitals

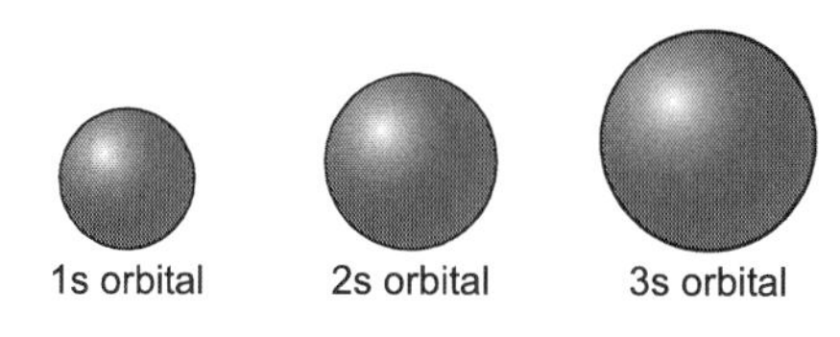

....

An atomic orbital is the volume in space where the probability of finding an electron is more than 90%. So in Figure 2.4 the s-orbitals shown are spherical in shape, the diameter of the sphere increasing as the value of *n* increases. At any instant in time there is approximately a 90% chance of finding the electron within the sphere.

Third Quantum number (also known as magnetic quantum number), symbol ***m***, relates to the orientation in space of the orbital. It is dependent on ℓ because *m* can take on any whole number value between -ℓ and +ℓ.

So if ℓ = 2 (labelled as a 'd' orbital)

m could have the value +2, +1, 0, -1, -2

So for ℓ = 2 there are five atomic orbitals.

The relationship between ℓ and *m* is shown in Table 2.2

Table 2.2: Relating ℓ and m

Sub-shell name	Possible value: ℓ	Possible value: *m*
1*s*	0	0
2*s*	0	0
2*p*	1	-1, 0,+1
3*s*	0	0
3*p*	1	-1, 0,+1
3*d*	2	-2,-1, 0,+1,+2

....

The relationship between *n*, ℓ and *m* can be summarised in the next activity.

Go online

Relating quantum numbers

Q2: Complete the table by filling the gaps with the numbers given below.

Value of n	Value of l	Value of m	Subshell name
1	0	0	
2	0 	 -1 0 +1	2s 2p
	0 1 2	0 -2 -1 0 +1 +2	3s 3p

-1 0 +1 | 0 | 3d | 1

1s | 3

....

2.4 Orbital shapes

Every atomic orbital can hold a maximum of two electrons and has its own shape, dictated by the quantum numbers.

s orbitals: are spherical with size and energy increasing as the value of n increases (see Figure 2.4).

p orbitals: have a value of $\ell = 1$ and there are, therefore, three possible orientations in space, corresponding to m = -1, 0, +1. The three p-orbitals are **degenerate** (have the same energy as each other), and have the same shape, approxiately dumbbell shaped and at right angles to one another (Figure 2.5). Each orbital is defined as if it lies along a set of x, y, z axes. The $2p$ orbitals are thus $2p_x$, $2p_y$ and $2p_z$. The $3p$ orbitals would be the same shape but larger and at higher energy.

Figure 2.5: Shapes of p orbitals

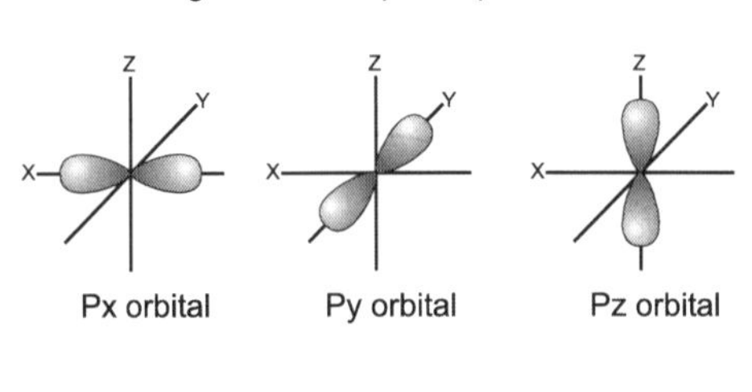

....

d orbitals: occur in five different orientations (Figure 2.6) corresponding to the third quantum number *m* = +2, +1, 0, -1, -2 and have labels which come from the complex mathematics of quantum mechanics. These orbitals are important in examination of the properties of transition metals. Note that d orbitals are degenerate (of equal energy) as well as p orbitals in an isolated atom. Three of the d orbitals lie between axes and two d orbitals lie along axes.

Figure 2.6: Shapes of d orbitals

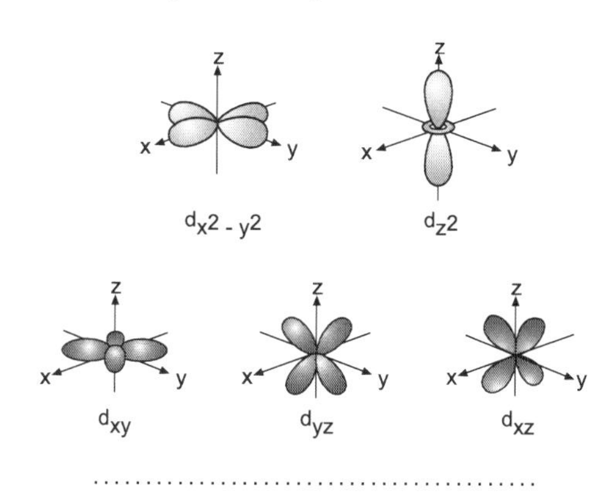

....

f orbitals: occur in seven different orientations relating to $\ell = 3$. The shapes are complex and do not concern us.

Each atomic orbital can hold a maximum of two electrons. Each electron in an orbital has a spin which causes it to behave like a tiny magnet. It can spin clockwise, represented as

[↑]

or anti-clockwise

[↓]

In any orbital containing two electrons they must be paired, with the spins opposed, sometimes represented as

[↑↓]

where the box represents the atomic orbital.

Figure 2.7: Electron spin

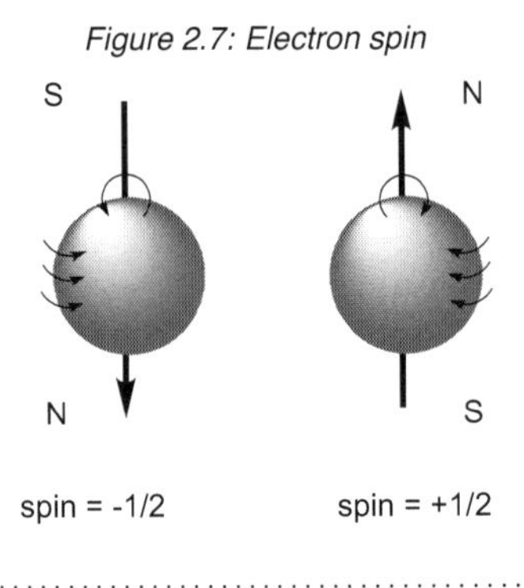

Fourth quantum number determines the direction of spin. It is therefore called the spin quantum number, ***s***. It has values of

$$+\frac{1}{2}\ or\ -\frac{1}{2}$$

(See Figure 2.7)

2.5 Electronic configurations

The arrangement of electrons in the energy levels and orbitals of an atom is called the electronic configuration. This can be expressed in three different ways.

1. Using quantum numbers

The four quantum numbers provide an address for the electron. No two electrons can have the same four quantum numbers, e.g. the three electrons in a lithium atom would have these addresses :

	Principal: ***n***	Second: ℓ	Third: ***m***	Fourth: ***s***
First electron	1	0	0	$+^1/_2$
Second electron	1	0	0	$-^1/_2$
Third electron	2	0	0	$+^1/_2$

2. Orbital box notation

Each orbital in an atom is represented by a box and each electron by an arrow. The boxes are filled in order of increasing energy. The orbital boxes for the first two energy levels are shown in Figure 2.8.

Figure 2.8: Orbital boxes

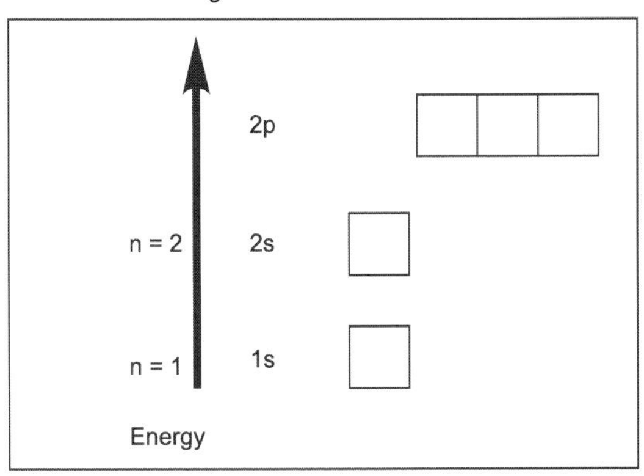

....

Using this notation, the electronic configuration for lithium (3 electrons) and carbon (6 electrons) is shown (Figure 2.9). Notice the presence of unpaired electrons.

Figure 2.9: Electron configuration of lithium and carbon

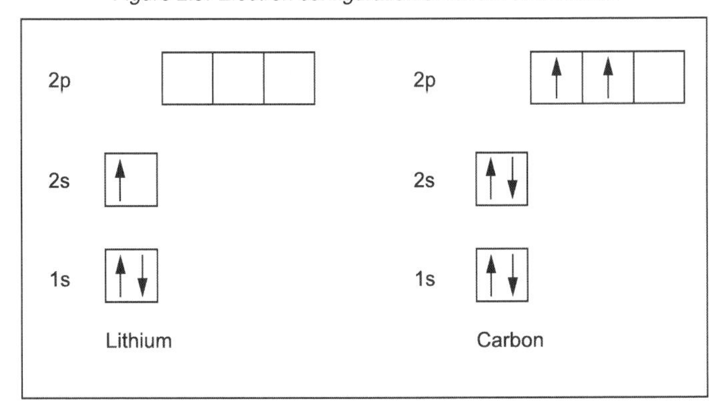

....

Note that you will sometimes come across half-headed arrows in questions. You can draw either full-headed or half-headed arrows when showing box notation.

3. Spectroscopic notation

This uses a shorthand representation of the arrangement of the electrons, so that the notation taken from Figure 2.9 for carbon would be:

$1s^2 2s^2 2p^2$

Notice that the designation $2p^2$ does not show which of the degenerate orbitals the electrons occupy.

Spectroscopic notations are sometimes made shorter by labelling the core of filled inner shells with the configuration of the preceding noble gas.

For example:

- Neon is $1s^2 2s^2 2p^6$
- Sodium is $1s^2 2s^2 2p^6 3s^1$
- Sodium can be written as $[Ne]3s^1$

Go online

Spectroscopic notation

Q3: Which of these is the spectroscopic notation for a lithium atom?

a) $1s^2\ 2s^2$
b) $2s^1\ 1s^2$
c) $1s^2\ 2s^1$
d) $2s^2 1s^1$

....

Q4: Which element has atoms with the spectroscopic designation $[Ar]4s^1$?

a) Hydrogen
b) Lithium
c) Chlorine
d) Potassium

....

Q5: How many electrons are there in the $2p$ sub-shell of the oxygen atom?

....

Q6: Which number would complete this spectroscopic notation for a nitrogen atom?

$1s^2\ 2s^2\ 2p^?$

....

Q7: Carbon has two **unpaired** electrons. How many **unpaired** electrons would boron have?

....

Q8: Which element is represented as $[Ne]\ 3s^2$?

a) Helium
b) Oxygen
c) Neon
d) Magnesium

....

Q9: Which of these represents the spectroscopic notation of a lithium **ion**? (Remember that a lithium ion has lost an electron to become Li^{+})

a) $1s^1$
b) $1s^2$
c) $1s^2\ 2s^1$
d) $1s^2\ 2s^2$

....

2.6 Writing orbital box notations

Multi-electron atoms can be represented in an orbital box diagram by applying three rules to determine the **ground state** electronic configuration.

Rule 1 'AUFBAU'	The **Aufbau principle** (From German 'building up'). When electrons are placed into orbitals the energy levels are filled up in order of increasing energy, e.g. in Figure 2.9 the 1*s* has filled before 2*s*.
Rule 2 'PAULI'	The **Pauli exclusion principle**. This states that an orbital cannot contain more than two electrons and they must have opposite spins, e.g. (↑) (↑)is **NOT** allowed.
Rule 3 'HUND'	**Hund's rule**. This states that when there are **degenerate** orbitals in a sub shell (as in 2*p*), electrons fill each one singly with spins parallel before pairing occurs. Thus carbon (Figure 2.9) has not paired up the two 2*p* electrons.

The relative energies corresponding to each orbital can be represented in order as far as 8*s*:

Figure 2.10: Aufbau diagram

1s
2s 2p
3s 3p 3d
4s 4p 4d 4f
5s 5p 5d 5f
6s 6p 6d
7s 7p
8s

. . . .

Following the arrows gives the order of orbital filling.

In orbital box notation:

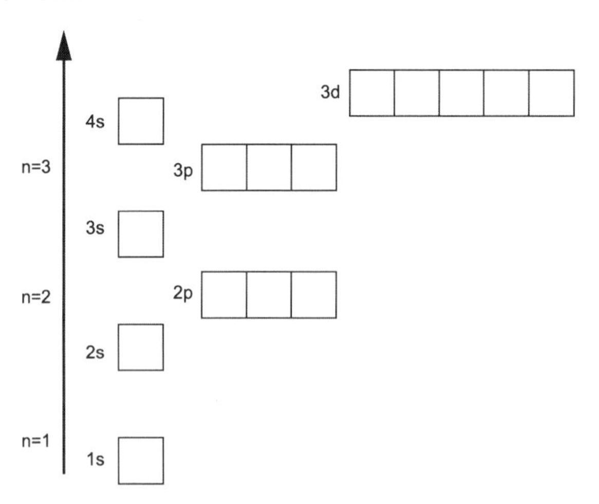

You may have noticed that the 4*s* orbital has been placed below the 3*d* in energy and think this is a mistake! It's no mistake, however. Although it is further from the nucleus in terms of space, it is lower in energy and gets filled up first.

Orbital box diagrams and spectroscopic notations can be worked out using the three rules.

Example : Working out the electronic configuration for an oxygen atom.

Rule 1	Start at the 1*s* level. Place electron 1 (of 8).
Rule 2	Pair up the second electron in 1*s*. Oppose spin. Repeat for 2*s* orbital. Four electrons are now placed.
Rule 3	Place electrons 5, 6, 7 into three degenerate 2*p*. Place electron 8 paired into any 2*p*.

Spectroscopic:

$$1s^2 2s^2 2p^4$$

Orbital box:

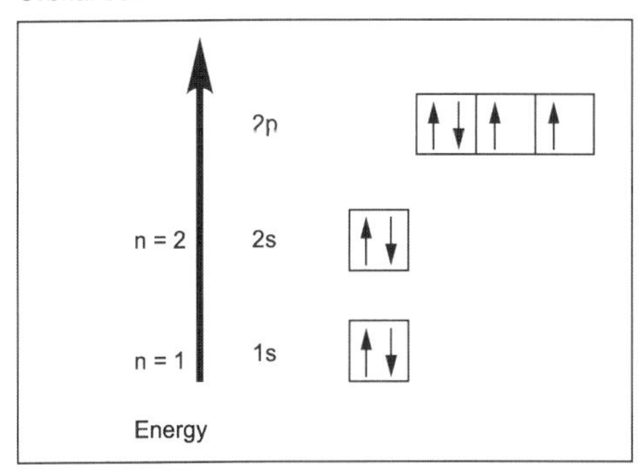

. . . .

Orbital box notation

Go online

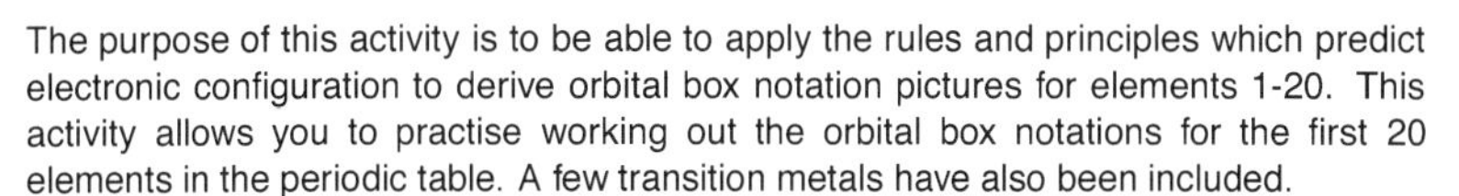
The purpose of this activity is to be able to apply the rules and principles which predict electronic configuration to derive orbital box notation pictures for elements 1-20. This activity allows you to practise working out the orbital box notations for the first 20 elements in the periodic table. A few transition metals have also been included.

Q10: To be able to apply the rules and principles which predict electronic configuration to derive orbital box notation pictures for elements 1-20. Using the orbital box below practise working out the orbital box notations for the first 20 elements in the periodic table.

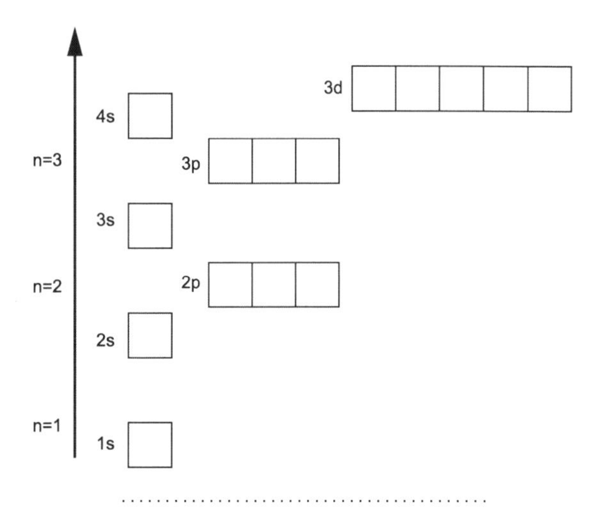

....

2.6.1 The Periodic Table

The structure of the Periodic Table depends upon the electronic configuration of the elements. Since the chemical properties of an element are dictated by the electrons in the outer shell, the Periodic Table relates configuration to properties. For example, fluorine has a configuration $1s^2$ $2s^2$ $2p^5$ and thus behaves like other Group Seven elements and is reactive because of its almost complete *p* sub-shell. The groups and periods of the Periodic Table are further organised into blocks. Fluorine is thus a *p*-block element, since the last sub-shell being filled is a *p* sub-shell.

Go online

Periodic table blocks

The purpose of this activity is to relate the structure of the Periodic Table to the electronic configuration of the elements.

H																	He
Li	Be											B	C	N	O	F	Ne
Na	Mg											Al	Si	P	S	Cl	Ar
K	Ca	Sc	Ti	V	Cr	Mn	Fe	Co	Ni	Cu	Zn	Ga	Ge	As	Se	Br	Kr
Rb	Sr	Y	Zr	Nb	Mb	Tc	Ru	Rh	Pd	Ag	Cd	In	Sn	Sb	Te	I	Xe
Cs	Ba	Lu	Hf	Ta	W	Re	Os	Ir	Pt	Au	Hg	Tl	Pb	Bi	Po	At	Rn
Fr	Ra	Lr	Rf	Db	Sg	Bh	Hs	Mt	Ds	Rg	Cn	Uut	Fl	Uup	Lv	Uus	Uuo

La	Ce	Pr	Nd	Pm	Sn	Eu	Gd	Tb	Dy	Ho	Er	Tm	Yb
Ac	Th	Pa	U	Np	Pu	Am	Cm	Bk	Cf	Es	Fm	Md	No

d f p s

Q11: Which areas represent elements which have *s*, *p*, *d* or *f* electrons in the outermost sub-shell?

....

Q12: Which block contains the noble gases (excluding Helium)?

a) *s*
b) *p*
c) *d*
d) *f*

....

Q13: Which block contains the most reactive metals?

a) *s*
b) *p*
c) *d*
d) *f*

....

Q14: What **one word** name is given to the *d*-block elements?

....

In each of the next four questions classify the element as being *s*, *p*, *d*, or *f* block.

Q15: Aluminium

....

Q16: $1s^2\ 2s^2\ 2p^6$

....

Q17: Scandium

....

Q18: $1s^2 2s^2\ 2p^6\ 3s^1$

....

....

2.7 Ionisation energy

The **ionisation energy** of an element is the energy required to remove one mole of electrons from one mole of the gaseous atoms. The second and subsequent ionisation energies refer to removal of further moles of electrons.

It can be represented as:

First Ionisation Energy:	$X(g) \rightarrow X^+(g) + e^-$
Second Ionisation Energy:	$X^+(g) \rightarrow X^{2+}(g) + e^-$

The variation in first ionisation energies for the first 36 elements (Figure 2.11) relates to the stability of the electronic configuration.

Figure 2.11: First ionisation energies of the elements

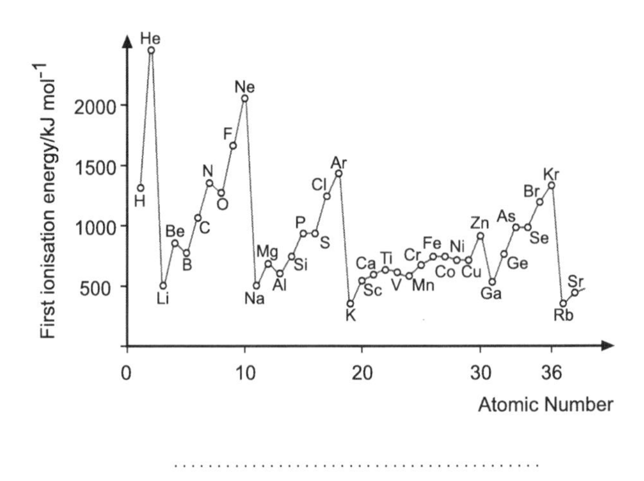

....

Refer to Figure 2.11 as you consider this question and answer set. Think it through.

Question	Answer
1. Which element group appears at the four peaks?	These are group **0**, noble gases.
2. Would the ionisation be easy or difficult?	**Difficult** as it is a high value.
3. Are these elements reactive or unreactive?	**Unreactive** noble gases.
4. How many electrons in the outer shell?	All are full, usually with **8** electrons.

Conclusion: the ionisation energy evidence supports the electronic configuration theory.

Ionisation energy evidence

Go online

Q19: Which group number has elements at the four lowest points?

....

Q20: Would the ionisation be relatively easy or difficult?

....

Q21: Are these elements reactive or unreactive?

....

Q22: How many electrons in the outer shell?

....

Conclusion: the ionisation energy evidence supports the electronic configuration theory.

There are two further features of Figure 2.11 which deserve some attention.

Feature 1: There is a slight dip from beryllium to boron.

Beryllium is:

$$\boxed{1s^2 2s^2}$$

and boron is:

$$\boxed{1s^2 2s^2 2p^1}$$

Beryllium has a full sub-shell and is more stable. Boron has a single 2*p* electron and is less stable.

Feature 2: There is a dip in the middle of the *p*-block from nitrogen to oxygen, see Figure 2.11.

Nitrogen is:

$$1s^2 2s^2 2p^3$$

and oxygen is:

$$1s^2 2s^2 2p^4$$

In orbital box notation:

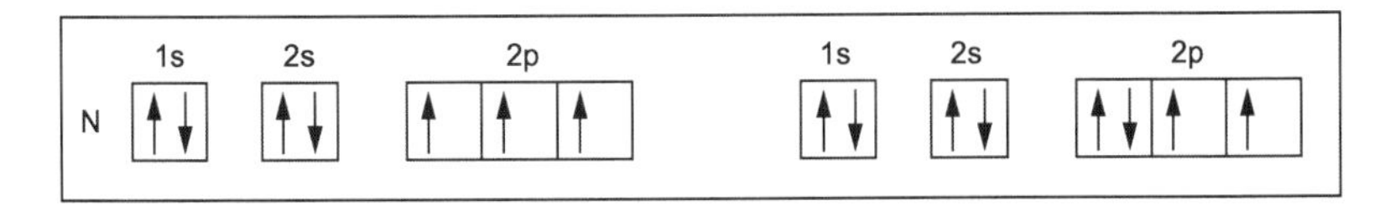

Half filled sub-shells are relatively stable and it is easier to remove the fourth **'p'** electron from the 2p shell of oxygen.

Look at Figure 2.11 again and find another part of the graph that can be explained in the same way as features **1** and **2**.

The first, second and successive ionisation energies also provide evidence of stability which can be explained by considering the electronic configuration.

Element	1st Ionisation Energy (kJ mol^{-1})	2nd Ionisation Energy (kJ mol^{-1})
Na	496	4562
Mg	738	1451

....

Go online

First and second ionisation energies

Ionisation energy information provides valuable evidence for the electronic configuration of atoms and the position of elements in the Periodic Table.

Q23: Complete the paragraph by putting the correct words from the following list in place:

- closer;
- complete;
- higher;
- magnesium;
- one electron;
- sodium;
- stronger;
- two electrons.

Sodium has .... in its outer shell whereas magnesium has .... in its outer shell. The first ionisation energy of magnesium is .... than that of sodium since magnesium has 12 protons in its nucleus and therefore has a higher nuclear charge and a .... attraction for the outer electrons. However, the second ionisation energy of .... is higher than that of .... since the electrons being removed come from a .... p subshell which is .... to the nucleus.

....

2.8 Summary

Summary

You should now be able to:

- relate spectral evidence to electron movements and ionisation energy;
- describe the four quantum numbers and relate these to atomic orbitals, their shape and relative energies;
- describe the four s, p and d atomic orbitals, their shapes and relative energies;
- relate the ionisation energies of elements to their electronic configuration, and therefore to their position in the Periodic Table;
- describe the electronic configuration of atoms 1-20 in spectroscopic notation;
- write electron configurations in spectroscopic notation.

2.9 Resources

- Royal Society of Chemistry (http://www.rsc.org)
- SciComm video (https://youtu.be/K-jNgq16jEY) - 3D models of orbitals.

2.10 End of topic test

Go online

End of Topic 2 test

The end of topic test for *Atomic orbitals, electronic configurations and the Periodic Table*

Q24: "Each atomic orbital can hold a maximum of only two electrons."
This is a statement of:

a) the Pauli exclusion principle.
b) Hund's rule.
c) the aufbau principle.
d) the Heisenberg uncertainty principle.

....

Q25: Which element has this spectroscopic notation?
[Ne] $3s^2$ $3p^5$

a) Neon
b) Boron
c) Nitrogen
d) Chlorine

....

Q26: How many quantum numbers are necessary to identify an electron in an atomic orbital?

a) 1
b) 2
c) 3
d) 4

....

Q27: Which of these electron arrangements breaks Hund's rule?

a)	↑↓	↑↓	↑	↑	
b)	↑↓	↑↓	↑↓	↑	↑
c)	↑↓	↑↓	↑↓	↑	
d)	↑↓	↑↓	↑	↑	↑

....

Q28: In the emission spectra of hydrogen, how many lines are produced by electron transitions involving only the three lowest energy levels?

a) 1
b) 2
c) 3
d) 4

....

Q29: The first and second ionisation energies of boron are 801 and 2427 kJ mol $^{-1}$ respectively.
This means that for one mole of gaseous boron 3228 kJ of energy:

a) would be needed to remove 2 moles of 2s electrons.
b) would be needed to remove 1 mole of 2p electrons and 1 mole of 2s electrons.
c) would be released when 1 mole of 2p electrons and 1 mole of 2s electrons are removed.
d) would be released when 2 moles of 2s electrons are removed.

....

Q30: The second ionisation energy of magnesium can be represented by:

a) $Mg^{+}(g) \rightarrow Mg^{2+}(g) + e^{-}$
b) $Mg(s) \rightarrow Mg^{2+}(g) + 2e^{-}$
c) $Mg(g) \rightarrow Mg^{2+}(g) + 2e^{-}$
d) $Mg^{2+}(g) \rightarrow Mg^{3+}(g) + e_{-}$

....

Q31: What is the total number of electrons which may occupy a p sub-shell and remain unpaired?

....

Q32: How many unpaired electrons are there in a fluorine atom?

....

Q33: What is the maximum number of quantum numbers which can be the same for any two electrons in an atom?

....

Q34: What word is used to describe the orbitals within the p, d or f sub-shells which have the same energy?

....

Q35: Which element has atoms with the same spectroscopic notation as a Calcium ion (Ca^{2+})?

....

Q36: Explain why the lines in an emission spectrum become closer and closer together as they converge towards the high energy end of the spectrum for an element.

....

Q37: The first ionisation energies for the p-block elements aluminium to argon follow an upward trend, with the exception of phosphorus.
Explain this in terms of the electronic configurations of phosphorus and sulfur.

....

....

Topic 3

Shapes of molecules and polyatomic ions

Contents

Prerequisite knowledge

Before you begin this topic, you should know:

- *types of bonding including metallic, non-polar covalent (pure covalent), polar covalent and ionic bonding. (Higher Chemistry, Unit 1, Chemical Changes and Structure) (Metallic bonding is not considered in this topic).*
- *that ionic and polar covalent bonding does not exist in elements. (Higher and National 5 Chemistry, Unit 1, Chemical Changes and Structure).*
- *electronegativity (attraction of an element for bonding electrons) (Higher Chemistry, Unit 1, Chemical Changes and Structure) values are found in the SQA data booklet.*
- *that the difference in electronegativity values of elements gives an indication to the likely type of bonding between atoms of different elements. (Higher Chemistry Unit 1 Chemical Changes and Structure).*

Learning objectives

By the end of this topic, you should be able to:

- *explain that covalent bonding involves the sharing of electrons and can describe this through the use of Lewis electron dot diagrams;*

- *predict the shape of molecules and polyatomic ions through consideration of bonding pairs and non-bonding pairs and the repulsion between them;*
- *understand the decreasing strength of the degree of repulsion from lone-pair/lone-pair to non-bonding/bonding pair to bonding pair/bonding pair.*

3.1 Covalent bonding

A covalent bond is formed when atomic orbitals overlap to form a molecular orbital. When a molecular orbital is formed, it creates a covalent bond. Pure covalent bonding (non-polar) and ionic bonding can be considered to be at opposite ends of a bonding continuum. Polar covalent bonding sits between these two extremes.

Valence shell electron pair repulsion (VSEPR) theory does not provide an accurate description of the actual molecular orbitals in a molecule. However, the shapes of molecules and polyatomic ions predicted are usually quite accurate.

The figure below shows the electrostatic forces in a hydrogen molecule. The positively charged nuclei will repel each other, as will the negatively charged electrons; but these forces are more than balanced by the attraction between the nuclei and electrons.

Figure 3.1: Attractive and repulsive forces in a hydrogen molecule

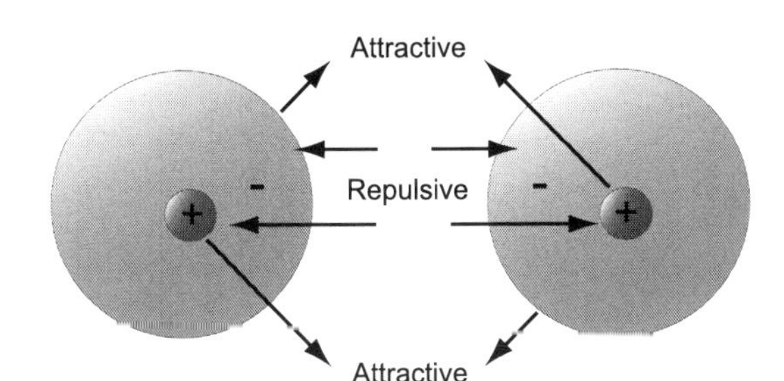

....

As the two Hydrogen atoms approach each other their atomic orbitals overlap and merge to form a molecular orbital forming a covalent bond between the two atoms.

The following graph relates distance between the atoms and potential energy.

Figure 3.2: Covalent bond formation

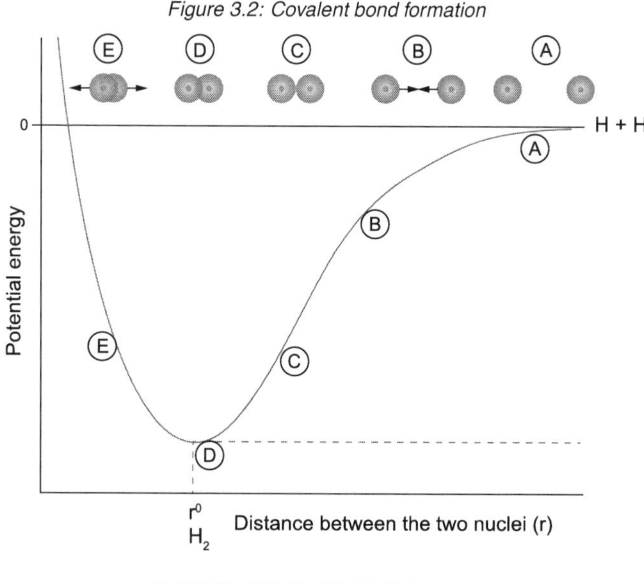

....

At the right hand side of the graph (the start) the two hydrogen atoms are far apart and not interacting (point A on the graph). As the hydrogen atoms approach each other, the nucleus of one atom attracts the neighbouring electron of the other atom and the potential energy drops. At the point when the two atoms become too close and the nuclei repel, the potential energy rises sharply. At point D on the graph a stable situation is reached where the attraction and repulsion are balanced and the lowest potential energy is reached. This is the point where the atomic orbitals have formed a new molecular orbital and the atoms are a bond length apart (r_0). In order to break these atoms apart the equivalent energy released in making the bond would have to be replaced. The quantity of energy required to break a mole of these bonds is known as the bond enthalpy.

Q1: At which point is the system most stable?

....

Q2: What does the symbol r° represent?

....

Q3: Would the progress from point A to point C be exothermic or endothermic?

....

Q4: The value of r° on this graph is 74 pm. Use the data book to look up the covalent radius for hydrogen and explain the apparent difference.

....

Q5: Use the information in the last question to predict the bond length in a molecule of hydrogen chloride.

....

3.2 Dative covalent bonds

When a covalent bond is formed two atomic orbitals join together to form a molecular orbital. Usually both atomic orbitals are half-filled before they join together. However sometimes one of the atoms can provide both the electrons that form the covalent bond and this is called a **Dative** Covalent bond. This bond is exactly the same as all other covalent bonds differing only in its formation. An example of this would be when an ammonium ion is formed in solution when an ammonia molecule picks up a hydrogen ion. The hydrogen ion (H^+) has no electrons and therefore cannot contribute an electron to the covalent bond. Both the electrons come from the lone pair on the nitrogen atom in the ammonia molecule. The dative bond is often arrowed.

$NH_3(aq) + H^+(aq) \rightarrow NH_4^+(aq)$

In a Lewis electron dot and cross diagram:

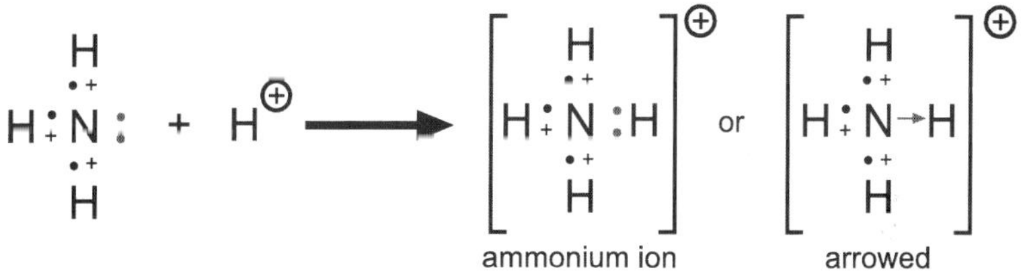

ammonium ion

or

arrowed

The dative bond is sometimes 'arrowed'.

3.3 Lewis diagrams

Lewis electron dot or dot and cross diagrams (named after American chemist G.N.Lewis) are used to represent bonding and non-bonding pairs in molecules and polyatomic ions.

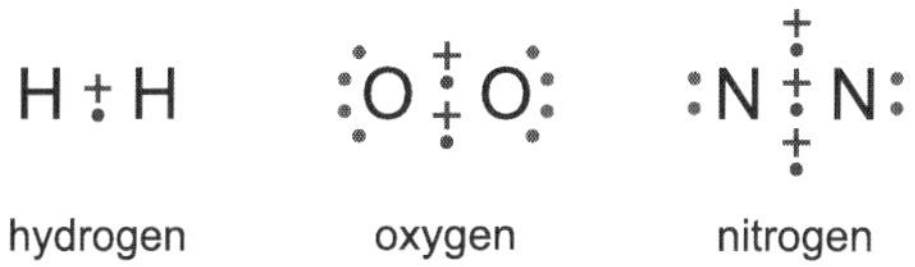

hydrogen oxygen nitrogen

Lewis electron dot diagrams

Dots can be used to represent all the electrons or you can use dots to represent only the electrons from one atom and crosses to represent electrons from the other atom. The above diagrams show that there is a single covalent bond between the hydrogen atoms, a double bond (2 pairs of electrons) between the oxygen atoms and a triple bond (3 pairs of electrons) between the nitrogen atoms.

Oxygen and nitrogen both have non-bonding electrons which are known as lone pairs (as shown by the outer paired dots). These have an influence on the chemistry of these molecules.

Go online

Resonance structures

Note that resonance structures are not required knowledge and students would not have to draw these from scratch. However, it is entirely possible that resonance structures may be included in the future in a problem solving/skills context.

The Lewis electron dot diagram for ozone O_3 shows there are 6 outer electrons from each oxygen atom giving a total of 18 electrons. There are 2 possible ways to draw the Lewis electron dot diagram for this molecule shown in the figure below. The two different forms are known as resonance structures. The actual structure of the ozone molecule is a hybrid of the two resonance structures also shown in the diagram below.

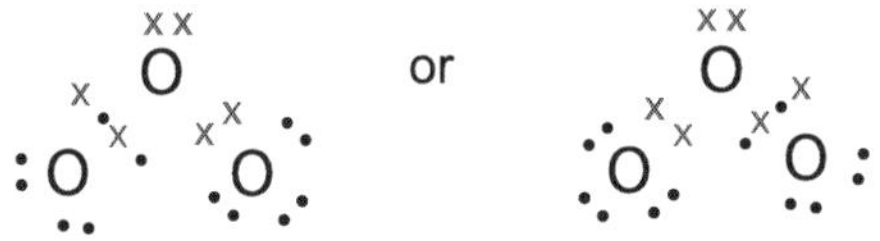

O or O

more easily drawn as:

Another structure with more than one resonance structure is the carbonate ion CO_3^{2-}. There are three different resonance structures for this polyatomic ion. See below.

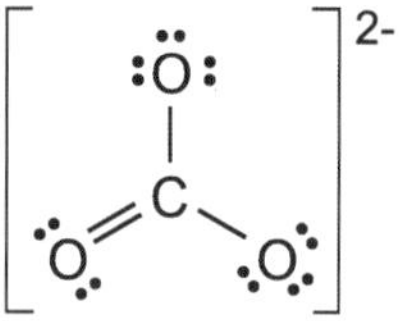

2-

Q6: Which molecule could have this Lewis electron dot diagram?

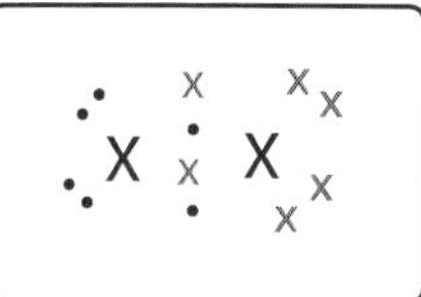

a) Cl_2
b) HCl
c) N_2
d) O_2

....

Q7: Which molecule could be represented by this dot and cross diagram?

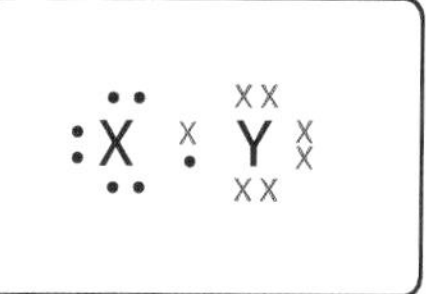

a) HCl
b) FCl
c) F_2
d) Cl_2

....

Q8: Draw a Lewis electron dot diagram to show the electrons in: a) methane and b) CO_2.

....

Q9: Carbon monoxide has a structure which contains a double bond and a dative covalent bond from oxygen to carbon. Draw this structure showing the dative bond as an arrow.

....

3.4 Shapes of molecules and polyatomic ions

Shapes of molecules and polyatomic ions can be predicted by first working out the number of outer electron pairs around the central atom and then dividing them into bonding and non-bonding (lone) pairs. The shape adopted by the molecule or polyatomic ion is one where these electron pairs can be as far apart as possible minimising repulsion between them.

Electron pairs = number of electrons on centre + number of bonded atoms/2.

If we apply this to a molecule of ammonia NH_3:

5 outer electrons on the central nitrogen atom (electron arrangement 2, 5) + 3 hydrogen atoms bonded.

5+3 (8)/2 = 4 (3 bonded and 1 lone pair)

Total number of electron pairs	Shape
2	Linear
3	Trigonal
4	Tetrahedral
5	Trigonal bipyramidal
6	Octahedral

3.5 Examples of molecules with different shapes

Two bonding pairs ($BeCl_2$)

Beryllium is in group two and therefore has two outer electrons. The two Cl atoms contribute one electron each giving four electrons in two electron pairs. As there are two Cl atoms bonded to the Be, these two electron pairs are bonding electrons and $BeCl_2$ will be a *linear* molecule with bond angles equal to 180°.

Cl - Be - Cl

Go online

Three pairs of electrons

Three bonding pairs ($BCl_3(g)$)

Boron is in group 3 and therefore has three electrons in the outer shell. As above the three Cl atoms provide one electron each giving six electrons in three electron pairs. All three electron pairs are involved in bonding and are therefore bonding electrons. No lone pairs exist on Boron. The $BeCl_3$ will have a *trigonal (trigonal planar)* shape with all four atoms in the same plane.

When this idea is extended to three pairs of electrons as in BCl_3 the molecule is flat with an angle of 120° and is described as trigonal planar.

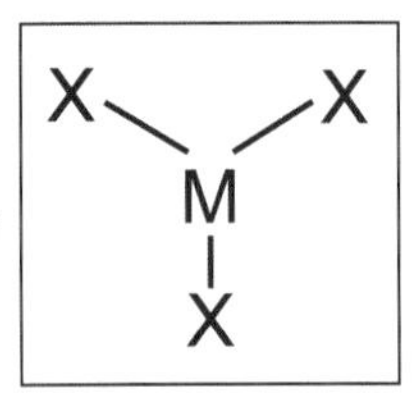

. . . .

Four bonding pairs (CH_4)

a) Perfect *tetrahedron* with bond angles of 109.5°.

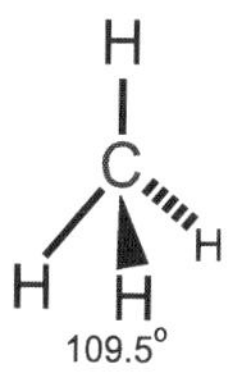

b) Four pairs of electrons with three bonding pairs and one lone pair (NH_3)

Nitrogen is in group five and therefore has five outer electrons. Each H atom provides one electron giving a total of eight electrons (four pairs) around the N atom. Three of the pairs are involved in bonding to the hydrogen atoms leaving one pair as a non-bonding pair (lone pair). Repulsion between a lone pair and a bonding pair is greater than repulsion between bonding pairs. This means the lone pair on the N atom pushes the three nitrogen hydrogen bonds closer together resulting in a slightly smaller bond angle of 107°. This would be described as a ***trigonal pyramidal*** molecule.

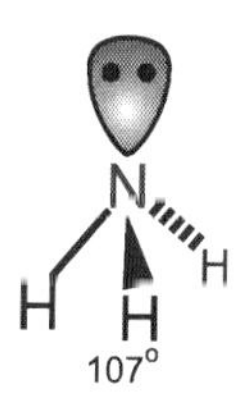

c) Four pairs of electrons with two bonding pairs and two lone pairs (H_2O)

Oxygen is in group six and therefore has six electrons in the outer shell. Each hydrogen atom contributes one electron making a total of eight (four pairs of) electrons around the central O atom. Two of the pairs are involved in bonding with the hydrogen atoms leaving two pairs as non-bonding (lone pairs). The two lone pairs push the two O-H bonds closer together due to greater repulsion between lone pairs giving a bond angle of 104.5°. The shape of this molecule is ***angular***.

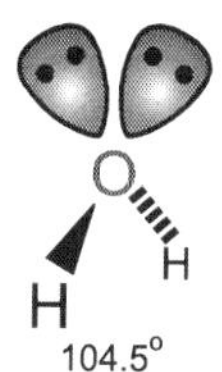

Go online

Five pairs of electrons

Five bonding pairs ($PCl_5(g)$)

Phosphorus has five outer electrons and each of the five chlorine atoms provides one electron giving a total of ten electrons (five pairs). All electron pairs are bonding pairs involved in P-Cl bonds. The shape of this molecule is *trigonal bipyramidal.*

....

When five pairs of electrons are involved the shape is said to be trigonal bipyramid with angles of 120°, 90° and 180°. An example is PCl_5.

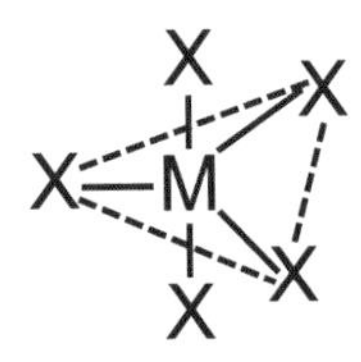

Remember that one or even two of these sites could be occupied by electrons only and the molecule shape would be changed from the trigonal bipyramid.

Go online

Six pairs of electrons

Six bonding pairs ($SF_6(g)$)

Sulfur has six outer electrons and each of the six fluorine atoms provides one electron giving a total of 12 electrons (six pairs). All electron pairs are bonding pairs involved in S-F bonds. The shape of this molecule is *octahedral.*

Sulfur hexafluoride

Six pairs of electrons is the highest number we are likely to encounter. For example in SF_6. Bond angles are 90°. Such a structure is described as octahedral. Octahedral geometry is relatively common.

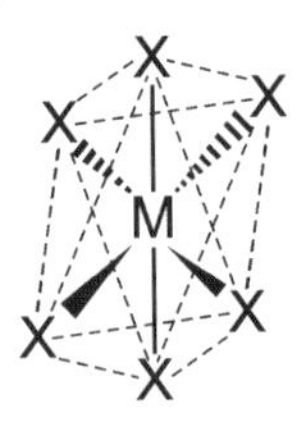

One other common structure worth mentioning is that resulting from an octahedral arrangement which involves two lone pairs.

The iodine tetrachloride negative ion

Six pairs of electrons with two of these as lone pairs results in their "repulsive power" keeping those two the furthest apart. For example in ICl_4^-. Bond angles are 90°. Such a structure is described as *square planar.*

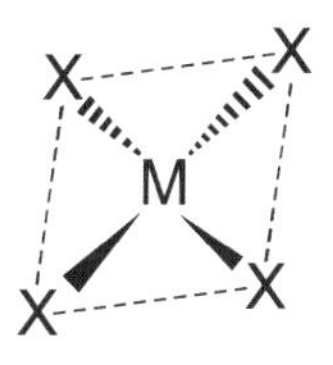

The negative charge in the iodine tetrachloride negative ion adds one electron to the total number.

A lone pair of non bonding electrons is more repulsive than a bonded pair. The different strength of electron pair repulsion accounts for slight deviations from the expected bond angles in a number of molecules.

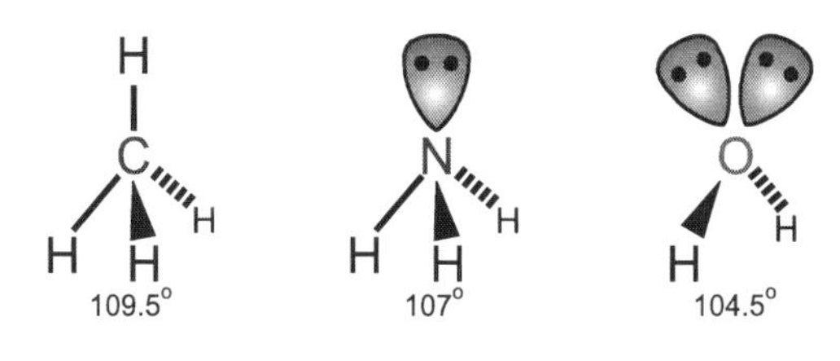

. . . .

Summary of shapes of covalent molecules

Go online

Q10:

Number of Electron Pairs	Arrangement	Angle(s) in degrees	Example
2	linear		
3			BF_3
4	tetrahedral		
5		90, 120, 180	
6			SF_6

Word Bank

180	trigonal planar	109.5	CH_4	90
PCl_5	octahedral	$BeCl_2$	120	trigonal bipyramidal

. . . .

Q11: Which of these molecules has a non-bonded electron pair on the central atom?

a) PF_3
b) BF_3
c) $BeCl_2$
d) H_2

. . . .

Q12: What is the likely structure of an antimony (V) chloride molecule?

a) Linear
b) Tetrahedral
c) Trigonal bipyramid
d) Octahedral

....

Q13: Which of these would have a bond angle greater than 109.5°?

a) CCl_4
b) NH_3
c) SCl_2
d) BeF_2

....

Q14: Sulfur has six outer shell electrons. Draw a diagram to show the SF_5^- ion structure (the negative ion).

....

Q15: What shape describes the arrangement of the electron pairs you have just drawn?

....

Q16: Suggest a name for the shape of this **molecule** (remember to ignore any lone pairs).

....

....

3.6 Summary

Summary

You should now be able to:

- explain that covalent bonding involves the sharing of electrons and can describe this through the use of Lewis electron dot diagrams;
- predict the shape of molecules and polyatomic ions through consideration of bonding pairs and non-bonding pairs and the repulsion between them;
- understand the decreasing strength of the degree of repulsion from lone-pair/lone-pair to non-bonding/bonding pair to bonding pair/bonding pair.

3.7 Resources

- Chemical Education Digital Library (http://bit.ly/29Jikva)
- Animated Molecules (http://bit.ly/29XBA76)
- Chemistry Pages http://bit.ly/29Sgu9O

3.8 End of topic test

Go online

End of Topic 3 test

Q17: Which of these compounds has the greatest degree of ionic character?

a) Beryllium oxide
b) Beryllium sulfide
c) Magnesium oxide
d) Calcium oxide

....

Q18: Forming a dative covalent bond between the phosphorus of PH_3 and the boron in BF_3 involves:

a) phosphorus losing electrons to boron.
b) boron losing electrons to phosphorus.
c) reducing the number of electrons in the boron outer shell.
d) phosphorus donating both electrons of the bond to boron.

....

Q19: Which of these is a non-linear molecule?

a) CO
b) CO_2
c) H_2S
d) $BeCl_2$

....

Q20: What change occurs in the three-dimensional arrangement of bonds around the boron in this reaction?

$BF_3 + F^- \rightarrow BF_4^-$

a) Trigonal planar to pyramidal
b) Trigonal planar to tetrahedral
c) Pyramidal to tetrahedral
d) Pyramidal to square planar

....

Q21: The bond angle in a molecule of ammonia is:

a) 90°
b) 107°
c) 109.5°
d) 120°

....

Q22: Identify the shape of the F_2O molecule.

a) Linear
b) Angular
c) Trigonal planar
d) Trigonal pyramidal
e) Tetrahedral
f) Trigonal bipyramidal

....

Q23: Identify the shape of the nitrogen fluoride molecule.

a) Linear
b) Angular
c) Trigonal planar
d) Trigonal pyramidal
e) Tetrahedral
f) Trigonal bipyramidal

....

Q24: Identify the **two** three-dimensional shapes around the oxygen atom in this reaction:

$H_2O + H^+ \rightarrow H_3O^+$

a) Linear
b) Angular
c) Trigonal planar
d) Trigonal pyramidal
e) Tetrahedral
f) Trigonal bipyramidal

....

This diagram shows the outer electron arrangements in a polyatomic ion.

The oxygen atoms are labelled A, B, C, D.

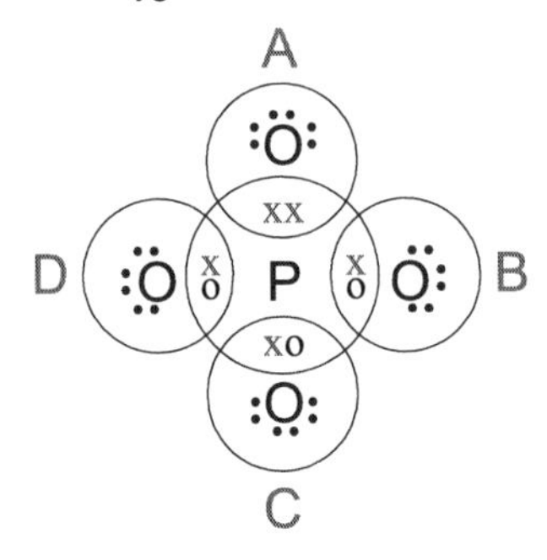

Q25: Which of these describes the shape of the ion?

a) Tetrahedral
b) Trigonal pyramidal
c) Square planar
d) Trigonal biplanar

....

Q26: What charge would this ion carry?

a) One negative
b) Two negative
c) Three negative
d) Five negative

....

Q27: Which of the oxygen atoms is attached by a dative covalent bond?

a) A
b) B
c) C
d) D

....

Q28: The image below shows a dot and cross diagram of how a dative covalent bond is formed in this reaction:

$BF_3 + F^- \rightarrow BF_4^-$

Identify the dative covalent bond.

The noble gas xenon can combine with fluorine under certain circumstances.

Xenon tetrafluoride is one possible product.

Q29: Which of the following shows a sketch of the xenon tetrafluoride molecule?

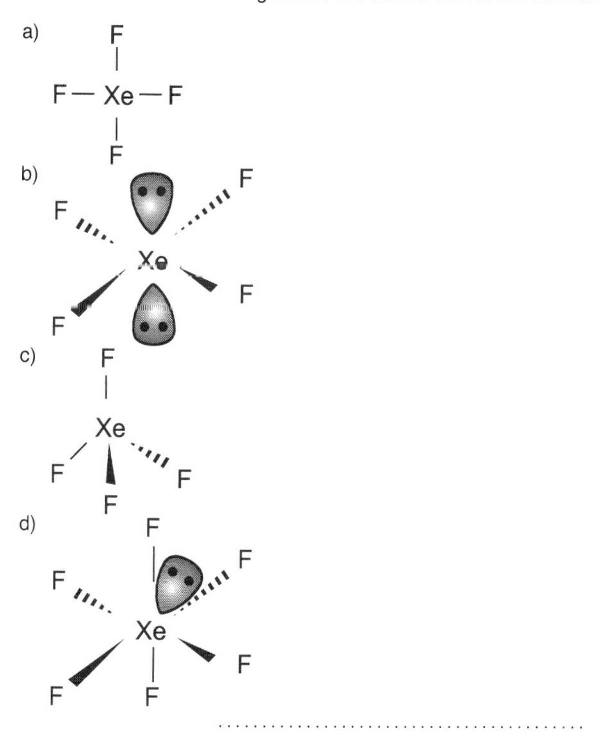

Q30: Name this shape.

Topic 4

Transition metals

Contents

Learning objectives

By the end of this topic, you should be able to:

- *understand and draw electronic configuration diagrams for transition metal atoms and ions;*
- *understand and explain any anomalies in the electronic configuration model;*
- *work out the oxidation state of transition metals and the oxidation number of transition metal ions;*
- *explain that changes in oxidation number show oxidation and reduction reactions;*
- *understand what allows a substance to be used as a ligand and how their classification and the coordination number are worked out;*
- *name complex ions according to IUPAC rules;*
- *explain what causes transition metal complexes to be coloured;*
- *understand UV and visible absorption spectroscopy of transition metal complexes;*
- *understand that transition metal complexes can be used in catalysis.*

4.1 Electronic configuration

Transition metals are found between groups 2 and 3 on the periodic table and are known as the d block elements. They have many important uses including piping, electrical wiring, coinage, construction and jewellery. Many have important biological uses and many are used as industrial catalysts.

The d block transition metals are metals with an incomplete d subshell in at least one of their ions. This gives transition metals their distinctive properties and we will be concentrating on the first row of transition metals from Scandium to Zinc.

As we go across the row from Scandium to Zinc the transition metals follow the aufbau principle, adding electrons to the subshells one at a time in order of their increasing energy, starting with the lowest. This must fit in with the electron arrangement given in the SQA data booklet.

Scandium has the electronic configuration $1s^2 2s^2 2p^6 3s^2 3p^6 3d^1 4s^2$

The 4s orbital has been filled before the 3d orbital due to being lower in energy.

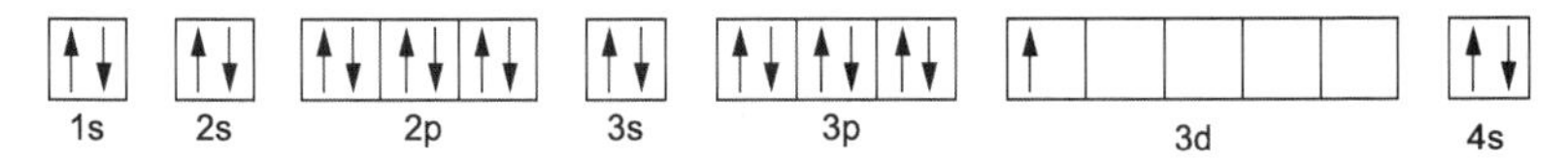

Electronic configuration of Scandium written in orbital box notation.

This can be shortened to [Ar] $3d^1$ $4s^2$ where [Ar] represents the s and p orbitals of the Argon core.

Go online

Orbital box notation

Q1:

Using the orbital box below practise working out the orbital box notations for the transition metals, Scandium, Titanium, Vanadium, Chromium, Manganese, Iron, Cobalt, Nickel, Copper and Zinc.

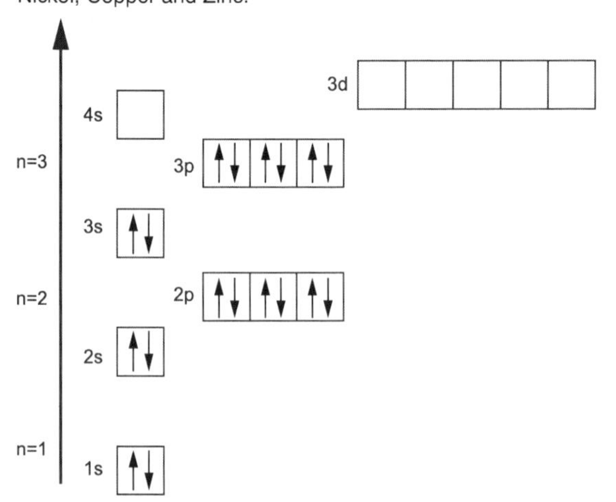

....

Copper and Chromium appear not to follow the aufbau principle (orbitals are filled in order of increasing energy).

Chromium [Ar] $3d^5$ $4s^1$

Copper [Ar] $3d^{10}$ $4s^1$

Half-filled or fully filled d orbitals have a special stability. ***However whenever transition metals form ions electrons are lost first from the outermost subshell the 4s.***

Electronic configuration of Co^{2+} is therefore [Ar] $3d^7$

Q2: Explain why Scandium and Zinc are often considered not to be transition metals.

....

Q3: Consider the electronic configurations of the Fe^{2+} and Fe^{3+} ions in terms of orbital box notation. Explain why Fe(III) compounds are more stable than Fe(II) compounds..

....

4.2 Oxidation states and oxidation numbers

The oxidation state is similar to the valency that an element has when it is part of a compound. Iron(II) chloride would normally be stated as having iron with a valency of 2, but it is actually more accurate to say that the iron is in an oxidation state (II) or has oxidation number +2.

Rules need to be followed when assigning an oxidation number to an element.

Rule No.	Rule
1.	Oxidation number of an uncombined element is 0.
2.	For ions containing single atoms (monoatomic) the oxidation number is the same as the charge on the ion. Example Na^+ and Cl^- the oxidation number would be +1 and -1 respectively.
3.	In most compounds oxygen has oxidation number -2.
4.	In most compounds hydrogen has the oxidation number +1. The exception is in metallic hydrides where it is -1.
5.	Fluorine always has oxidation number -1.
6.	The sum of all the oxidation numbers of all the atoms in a molecule or neutral compound must add up to 0.
7.	The sum of all the oxidation numbers of all the atoms in a polyatomic ion must add up to the charge on the ion.

4.2.1 Calculating an oxidation state

Oxidation number of Mn in MnO_4^-

We must apply rule 7 here where all the oxidation numbers of the atoms must add up to -1 (charge on the ion). Each oxygen atom has an oxidation number of -2 (rule 3) so the sum of the oxidation numbers on oxygen is 4 x-2 = -8. Therefore the oxidation number of Mn must be 7 (-8 + 7 = -1). Find the oxidation number for the transition metal in the following examples.

Q4: VO_2^+

....

Q5: CrO_4^{2-}

....

Q6: $VOCl_2$

....

Q7: Cr_2O_3

....

Q8: $K_2[Cr_2O_7]$

....

Q9: $Na_4[NiCl_6]$

....

Q10: $[FeO_4^{2-}]$

....

Q11: $K_2[MnO_4]$

....

Q12: $K_3[CoF_6]$

....

4.2.2 Multiple oxidation states

Transition metals may have more than one oxidation state in their compounds. Iron for example has the familiar oxidation states of (II) and (III). Copper is predominately in oxidation state (II) but can have an oxidation number of +1 in Cu_2O.

Transition metal compounds can exhibit different colours depending on the oxidation state of the metal. For example iron(II) compounds are often pale green and iron(III) compounds are yellow-orange. Iron(II) compounds are less stable than iron(III) since the iron(II) becomes slowly oxidised to iron(III).

The relative stabilities of the different oxidation states are determined by several factors including:

- the electronic structure (which influences ionisation energies and ionic radius);

- the type of bonding involved;
- the stereochemistry.

				7+					
			6+	(6+)	(6+)				
		5+							
	4+	4+		4+					
3+	(3+)	(3+)	3+	3+	3+	3+	(3+)		
	(2+)	(2+)	(2+)	2+	2+	2+	2+	2+	2+
								1+	
Sc	**Ti**	**V**	**Cr**	**Mn**	**Fe**	**Co**	**Ni**	**Cu**	**Zn**

Common oxidation states of the first transition metal series.

Less common oxidation states are shown in brackets.

4.2.3 Oxidation and reduction

OIL - Oxidation is a loss of electrons

This can also be also shown as an increase in the oxidation number of the transition metal.

RIG - Reduction is a gain of electrons

This can also be shown as a decrease in the oxidation number of the transition metal.

Determine if the conversion from VO^{2+} to VO_2^+ is oxidation or reduction.

VO^{2+}	VO_2^+
Overall charge of ion = +2	Overall charge of ion = +1
Oxidation number of O = -2	Oxidation number of O = 2x-2 = -4
Oxidation number of V = +2-(-2) = +4	Oxidation number of V = +1-(-4) = +5

The oxidation number of Vanadium has increased from +4 in VO^{2+} to +5 in VO_2^+ showing it has been oxidised.

Compounds containing metals in high oxidation states tend to be oxidising agents whereas those containing metals in low oxidation states tend to be reducing agents.

4.3 Ligands and transition metal complexes

Ligands are electron donors which are usually negative ions or molecules that have one or more non-bonding (lone) pairs of electrons. When these ligands surround a central transition metal ion they form a transition metal complex often called a coordination compound.

Chloride ion	Cyanide ion	Ammonia molecule	Water molecule
$:\ddot{Cl}:^{\ominus}$	$^{\ominus}:C\equiv N:$	$\ddot{N}H_3$	$H\ddot{O}H$

Monodentate ligands

These ligands are known as monodentate which means they donate one pair of electrons to the central transition metal ion i.e. form a dative bond. A bidentate ligand donates two pairs of electrons to the central transition metal ion and examples include the oxalate ion and 1, 2-diaminoethane (ethylene diamine abbreviated to 'en').

Oxalate	Ethylenediamine (abbreviated to 'en')
$^{\ominus}:\ddot{O}:$ $:\ddot{O}:^{\ominus}$ C—C $:O$ $O:$	$H_2\ddot{N}$—CH_2—CH_2—$\ddot{N}H_2$

Bidentate ligands

EDTA (ethylenediaminetetraacetate) is a hexadenate ligand as it has 6 non-bonding pairs of electrons which bond to the central transition metal ion. EDTA reacts with metal ions such as Ni^{2+} in a 1:1 ratio.

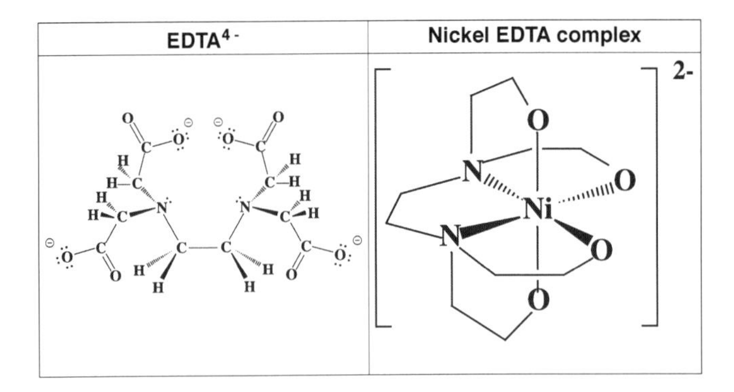

Nickel EDTA complex

4.3.1 Coordination number and shape of ligands

The **coordination number** of the central transition metal ion is the number of bonds from it to the ligands. The coordination number will determine the shape of the complex ion.

Coordination number	Shape	
2	Linear	X—M—X
4	Square planar	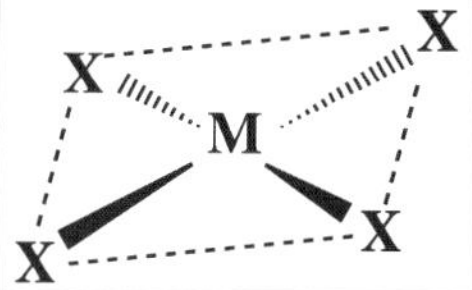
4	Tetrahedral	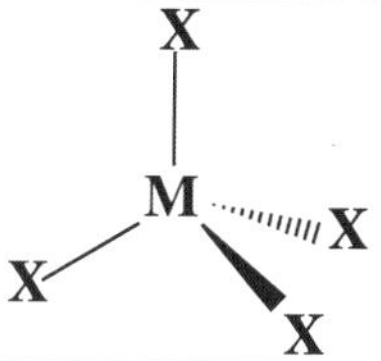
6	Octahedral	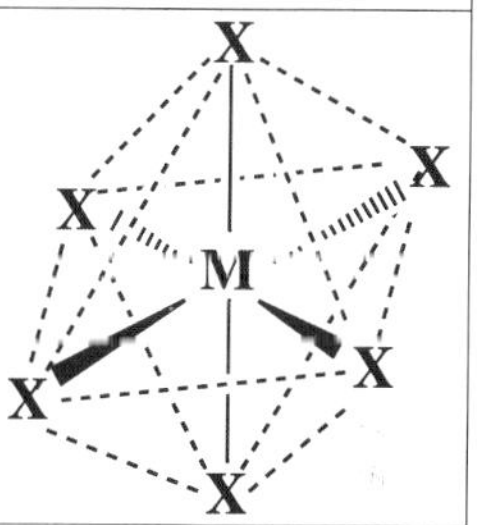

Coordination number and shape

4.3.2 Naming complexes

Transition metal complexes are named and written according to IUPAC rules.

Rule No.	Rule
1	The symbol of the metal is written first, followed by negative ligands, followed by neutral ligands.
2	Formula of the complex ion is enclosed within square brackets $[Fe(OH)_2(H_2O)_6]^+$ with the charge outside the square brackets.
3	Ligands are named in alphabetical order followed by the name of the metal and its oxidation state. If there is more than one of a ligand it is preceded by the prefix for the number di, tri, tetra etc.
4	If the ligand is a negative ion ending in -ide then in the complex name the ligand name changes to end in 'o'. Chloride become chloro and cyanide becomes cyano.
5	If the ligand is ammonia NH_3 it is named as ammine. Water as a ligand is named aqua.
6	If the complex is a negative ion overall the name of the complex ends in -ate. Cobaltate would be for a negative ion containing cobalt. However for copper cuprate is used and ferrate for iron.
7	If the complex is a salt the name of the positive ion precedes the name of the negative ion.

Some common ligands and their names:

- Ammonia, NH_3 (ammine)
- Bromide, Br^- (bromo)
- Carbon monoxide (carbonyl)
- Chloride, Cl^- (chloro)
- Cyanide, CN^- (cyano)
- Fluoride, F^- (fluoro)
- Hydroxide, OH^- (hydroxo)
- Iodine, I^- (iodo)
- Nitrite, NO_2^- (nitro if it bonds through the N and nitrito if it bonds through the O)
- Oxalate, $C_2O_4^{2-}$ (oxalato)
- Oxide, O^{2-} (oxo)
- Water, H_2O (aqua)

Examples of naming complex ions

$[Cu(H_2O)_4]^{2+}$ is named tetraaquacopper(II)

$[Co(NH_3)_6]^{2+}$ is named hexaamminecobalt(II).

$[Fe(CN)_6]^{4-}$ is named hexacyanoferrate(II).

If we have $K_3[Fe(CN)_6]$ this would be called potassium hexacyanoferrate(III).

$(K+)_3$ each K has a 1+ charge so three would contribute a 3+ charge. This means the negative ion from the complex will have an overall 3- charge. Each cyanide ion contributes a -1 charge so six of them would contribute a -6 charge. This means the oxidation state of iron would be 3+.

$(CN)_6$ = -6

Overall charge on negative ion = -3

Oxidation state of Fe = (-3 + -6) = +3

Naming transition metal complexes

Go online

Part 1

For each of the following complexes, write the correct name. Be very careful to spell each part of the name accurately (no capital letters) and don't put in spaces unless they are needed.

Q13: $[Co(H_2O)_6]Cl_2$

....

Q14: $Na[CrF_4]$

....

Q15: $K_4[Fe(CN)_6]$

....

Q16: $K_3[Fe(C_2O_4)_3]$

....

Part 2

For each of the following compounds what is the coordination number of the transition metal ion?

Q17: $Na[CrF_4]$

....

Q18: $K_3[Fe(C_2O_4)_3]$

....

Q19: $K_4[Fe(CN)_6]$

....

Q20: Predict the shape of the complex ion in the previous question.

....

Part 3

What is the correct structural formula for each of the following compounds?

Q21: Sodium tetrachloroplatinate(II)

a) $Na[PtCl_4]$
b) $Na_2[PtCl_4]$
c) $Na[Pt_2Cl_4]$
d) $Na_4[PtCl_4]$

....

Q22: Diaquadicyanocopper(II)

a) $(H_2O)_2(CN)_2Cu$
b) $[Cu(CN_2)(H_2O_2)]$
c) $Cu(CN)(H_2O)_2$
d) $[Cu(CN)_2(H_2O)_2]$

....

Q23: Pentaquachlorochromium(III) chloride

a) $[CrCl(H_2O)_5]Cl_2$
b) $[Cr(H_2O)_5]Cl_3$
c) $[CrCl(H_2O)_5]Cl_3$
d) $[CrCl_5(H_2O)_5]Cl$

....

Q24: Tetraamminedichlorocobalt(III) chloride

a) $[CoCl_2(NH_3)_4]Cl_3$
b) $[CoCl_2(NH_3)_4]Cl_2$
c) $[CoCl_2(NH_3)_4]Cl$
d) $[Co(NH_3)_4]Cl_3$

....

....

4.4 Colour in transition metal complexes

Several transition metal complexes are coloured including solutions of copper(II) compounds which are blue and solutions of nickel(II) complexes are green. To explain how these colours arise we need to look at the identity and oxidation state of the transition metal and the ligands attached in the complex.

White light is a complete spectrum ranging from 400 to 700nm known as the visible region of the electromagnetic spectrum. White light consists of all the colours of the rainbow. A complex appears coloured when some of this spectrum is absorbed and colourless when none is absorbed. If all the colours are absorbed the complex will appear black.

Transition metal complexes are able to absorb light due to the five degenerate d orbitals splitting in terms of energy. In a free transition metal ion (one without ligands) the five d orbitals in the 3d subshell are degenerate (equal in energy). On the formation of a complex for example $[Ni(H_2O)_6]^{2+}$ six water ligands surround the central nickel ion forming an octahedral shaped complex. The ligands approach the Nickel ion along the x,y and z axes. The electrons in d orbitals that lie along these axes (dz^2 and dx^2-y^2) will be repelled by electrons in the water ligand molecules. These orbitals now have higher energy than the three d orbitals that lie between the axes (dxy, dyz and dxz) and therefore the five d orbitals are no longer degenerate. This is called splitting of d orbitals and is different in octahedral complexes compared to tetrahedral and other shapes of complexes.

The energy difference between the different subsets of d orbitals depends on the ligand and its position in the **spectrochemical series** (series of order of ligand's ability to split the d orbitals).

$$CN^- > NH_3 > H_2O > OH^- > F^- > Cl^- > Br^- > I^-$$

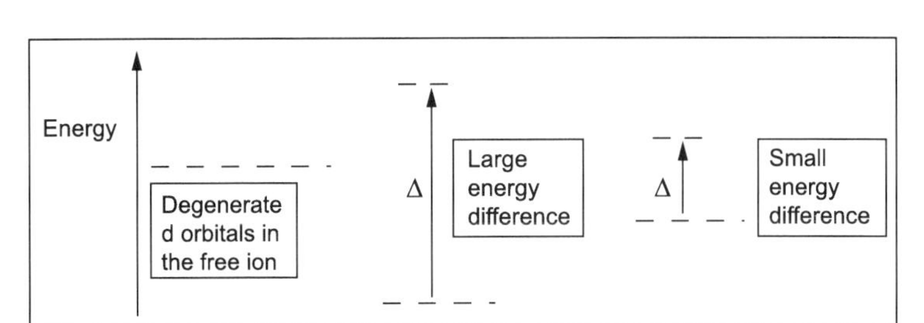

The difference in energy between the two subsets of d orbitals is known as the crystal field strength. This is given the symbol delta Δ.

Go online

Colour of transition metal compounds

Key point

Compounds are coloured because they absorb radiation from the visible part of the spectrum. The colour of a compound is that of the light which is not absorbed.

Part 1

White light consists of all the wavelengths of light in the visible spectrum combined. When compounds absorb radiation from the visible spectrum, the colour corresponding to this wavelength is removed from the white light and the colour that remains is the complementary colour. The colour wheel below shows complementary colours of light, i.e. if green light is absorbed then red light is transmitted.

Figure 4.1: Colour wheel

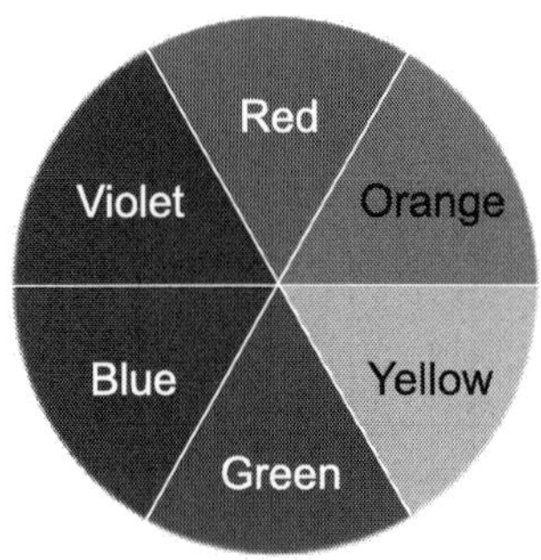

Red
Orange
Yellow
Green
Blue
Violet

. . . .

Q25: Complete the following passage using words from the following word list:

- blue;
- green;
- orange;
- red;
- violet;
- white:
- yellow.

If violet is absorbed, ... is transmitted.

If .... is absorbed, green is transmitted.

If orange is absorbed, ... is transmitted.

When all colours of light are present .... light is produced.

. . . .

Part 2

Figure 4.2: Absorption of colour by a complex

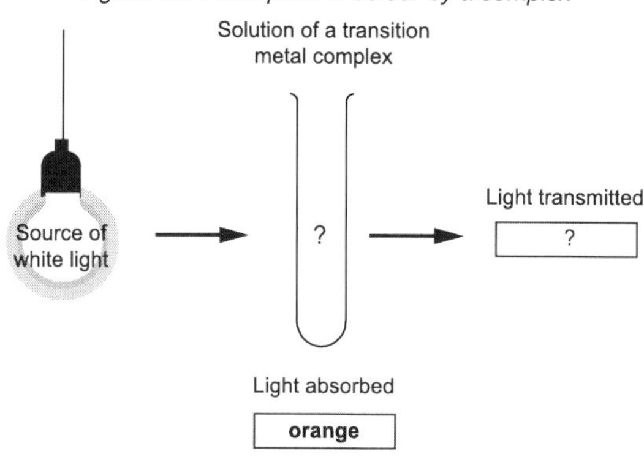

....

In the image above, white light is passed through a solution of a transition metal complex. Some visible light is absorbed.

Q26: What colour of light is transmitted?

....

Q27: What colour is the solution?

....

Q28: If violet light had been absorbed, what colour would the solution have appeared?

....

When one particular colour of light is absorbed, the colour remaining is the **complementary colour**. In other words the transmitted light is the complementary colour of the absorbed light.

....

4.5 UV and visible spectroscopy

Transition metal complexes absorb light due to the split in d orbitals. Electrons in the lower d orbitals can absorb energy and move to the higher energy d orbitals. If this energy absorbed in this d-d transition is in the visible region of the electromagnetic spectrum the colour of the transition metal complex will be the complementary colour of the colour absorbed.

The effects of d-d transitions can be studied using spectroscopy. If the absorbed energy is in the visible part of the electromagnetic spectrum (400-700nm) the complex will be coloured and visible spectroscopy would be used. If the absorbed energy is in the UV part of the electromagnetic spectrum (200-400nm) the transition metal complex will be colourless and UV spectroscopy will be used.

If the ligands surrounding the transition metal ion are strong field ligands (those that cause the greatest splitting of the d orbitals) d-d transitions are more likely to occur in the UV region of the electromagnetic spectrum. If the ligands are weak field ligands (those that split the d orbitals least) the energy absorbed is more likely to occur in the visible region of the electromagnetic spectrum. These complexes will be coloured.

A colorimeter fitted with coloured filters corresponding to certain wavelengths in the visible region can be used to measure the absorbance of coloured solutions. A filter of the complementary colour should be used.

Figure 4.3: Ultraviolet / visible spectrometer

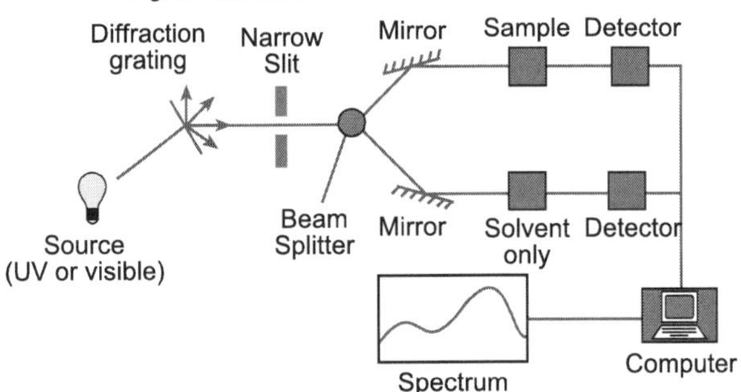

....

Samples are used in solution and are placed in a cell. Another identical cell containing the pure solvent is also placed in the machine. Radiation across the whole range is scanned continuously through both the sample solution and the pure solvent. The spectrometer compares the two beams. The difference is the light absorbed by the compound in the sample. This data is produced as a graph of wavelength against absorbance. An example is shown in the figure below.

Figure 4.4: Ti $^{3+}$(aq) visible spectrum

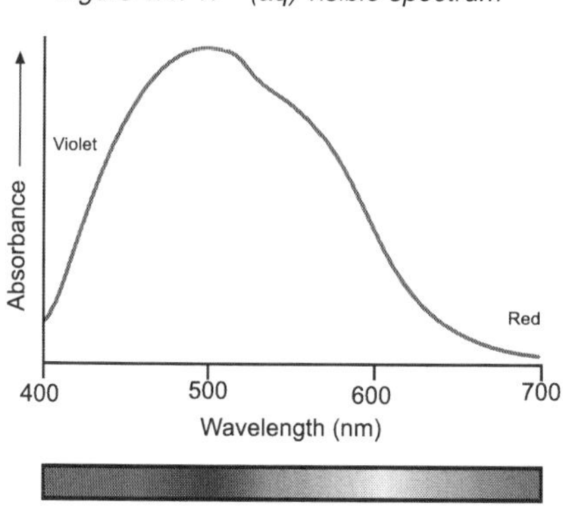

....

A UV spectrometer passes different wavelengths of UV light from 200 to 400nm through a sample and the quantity of UV light absorbed at different wavelengths is recorded. The intensity of the light absorbed at a given wavelength (especially the wavelength of maximum absorbance) is proportional to concentration, therefore UV / visible spectrum can also be used for quantitative analysis (colorimetry).

Explanation of colour in transition metal compounds

Go online

In this activity, three chromium(III) complexes will be considered. All are octahedral complexes which differ only in the nature of the ligands surrounding the central chromium(III) ion. The chromium(III) ion has a d^3 configuration. In an octahedral complex the d orbitals will be split and absorption of energy in the visible region can promote an electron from the lower to the higher level. When electrons are promoted to higher energy levels, the colour seen is entirely due to this absorption and the complementary colour being transmitted.

Note that no photon emission is involved as the electron falls back down to the lower energy level. This is a very common misconception by students who often confuse the two concepts. Generally emission will only occur during a flame test (or after a high energy electrical spark). If emission did occur, then the colour seen would be the same wavelength as the colour absorbed, i.e. not the complementary colour.

Figure 4.5: Splitting of d orbitals in an octahedral complex

Δ

Degenerate orbitals Split field

....

Figure 4.6: Absorption of a photon of light by an octahedral complex

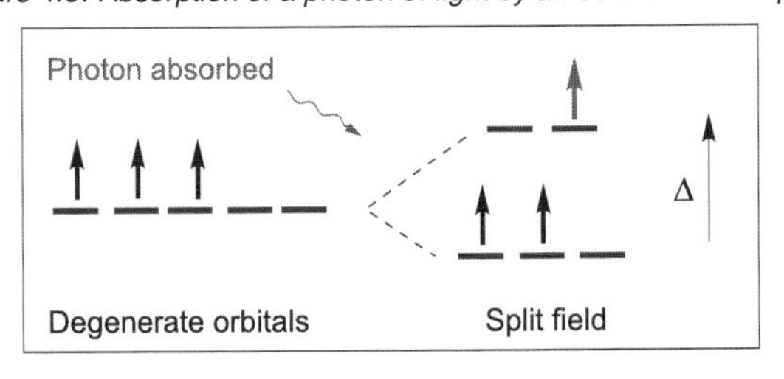

....

The hexachlorochromate(III) ion formula: $[CrCl_6]^{3-}$

Figure 4.7: Visible spectrum of hexachlorochromate(III) ion: note non linear scale on x axis

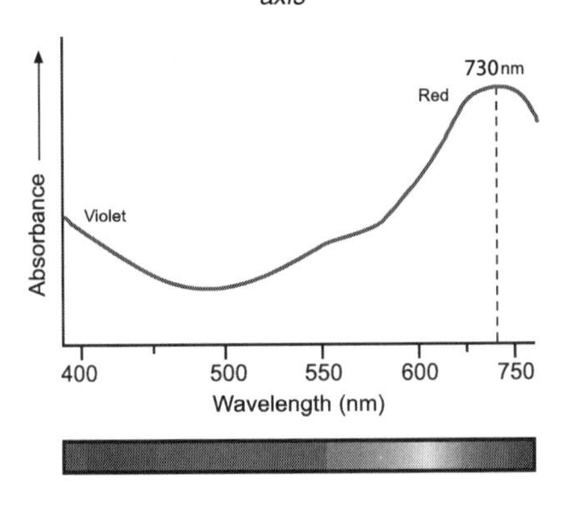

....

Q29: Use the wavelength of the most intense absorption in the visible spectrum (see the figure above) to calculate in kJ mol^{-1} the crystal field splitting (Δ) caused by the chloride ion (give your answer to one decimal place).

....

Q30: What colour would you predict for a solution containing $[CrCl_6]^{3-}$ ions?

a) Red
b) Blue
c) Violet
d) Yellow

....

The hexaaquachromium(III) ion formula: $[Cr(H_2O)_6]^{3+}$

Figure 4.8: Visible spectrum of hexaaquachromium(III) ion: note non linear scale on x axis

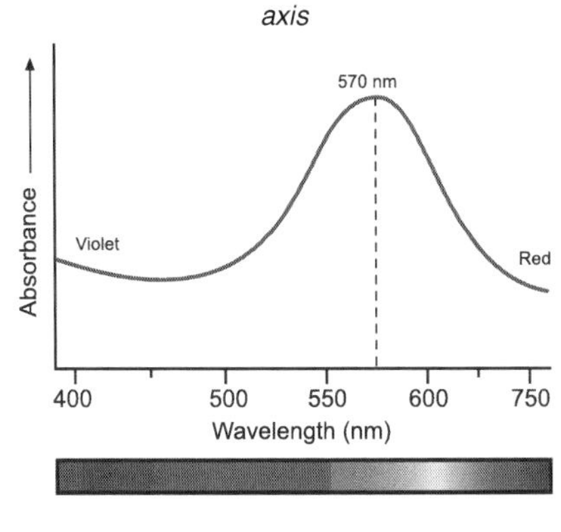

....

Q31: Use the wavelength of the most intense absorption in the visible spectrum (see the figure above) to calculate in kJ mol^{-1} the crystal field splitting (Δ) caused by the water ligand (give your answer to one decimal place).

....

Q32: What colour would you predict for a solution containing $[Cr(H_2O)_6]^{3+}$ ions ?

a) Red
b) Yellow
c) Violet
d) Green

....

The hexaamminechromium(III) ion formula: $[Cr(NH_3)_6]^{3+}$

Figure 4.9: Visible spectrum of hexaamminechromium(III) ion: note non linear scale on x axis

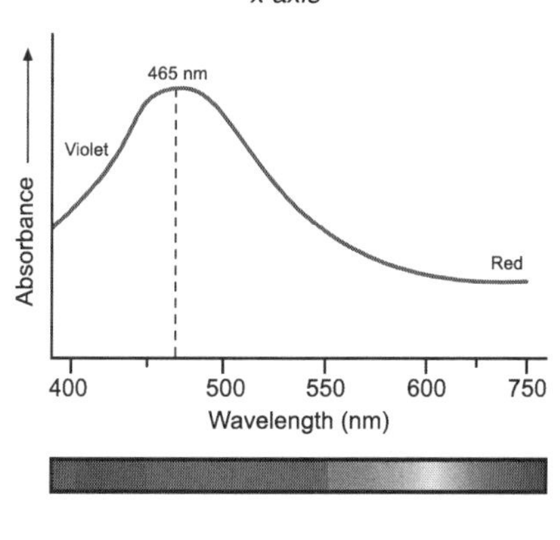

....

Q33: Use the wavelength of the most intense absorption in the visible spectrum (see the figure above) to calculate in kJ mol^{-1} the crystal field splitting (Δ) caused by the ammonia ligand (give your answer to one decimal place).

....

Q34: What colour would you predict for a solution containing $[Cr(NH_3)_6]^{3+}$ ions?

a) Red
b) Yellow
c) Violet
d) Green

....

Q35: The ligands can be placed in order of the crystal field splitting (Δ) with the ligand of lowest energy first. Which of the following shows the correct order?

a) $NH_3 < H_2O < Cl^-$
b) $NH_3 < Cl^- < H_2O$
c) $Cl^- < H_2O < NH_3$
d) $Cl^- < NH_3 < H_2O$

....

Key point

Different ligands produce different crystal field splittings and so complexes of the same metal ion with different ligands will have different colours.

....

4.6 Catalysis

Transition metals and their compounds are used as catalysts.

Transition Metal	Process
Iron	Haber Process production of ammonia
Platinum	Ostwald Process production of nitric acid
Platinum/Palladium/Rhodium	Catalytic converters
Nickel	Hardening of oil to make margarine
Vanadium	Contact Process production of sulfuric acid

These are examples of heterogeneous catalysts as they are in a different physical state to the reactants. Transition metals such as iron, copper, manganese, cobalt, nickel and chromium are essential for the effective catalytic activity of certain enzymes showing their importance in biological reactions.

Transition metals are thought to be able to act as catalysts due to atoms on the surface of the active sites forming weak bonds with the reactant molecules using partially filled or empty d orbitals forming intermediate complexes. This weakens the covalent bonds within the reactant molecule and since they are now held in a favourable position they are more likely to be attacked by molecules of the other reactant. This provides an alternative pathway with a lower activation energy increasing the rate of reaction.

Hydrogenation of ethene using a nickel catalyst

Go online

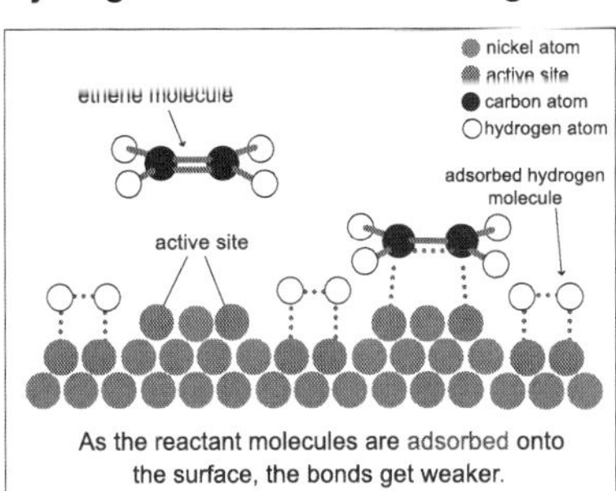

As the reactant molecules are adsorbed onto the surface, the bonds get weaker.

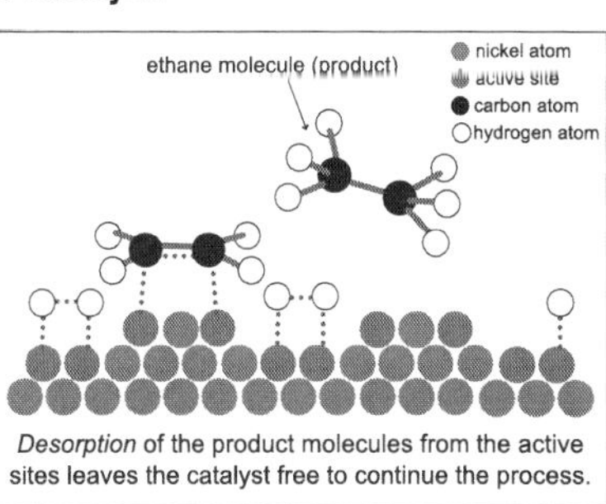

Desorption of the product molecules from the active sites leaves the catalyst free to continue the process.

Note: Adsorption is where something sticks to the surface.

Transition metals are also thought to be able to act as catalysts due to having variable oxidation states. This also allows the transition metal to provide an alternative pathway with a lower activation energy.

Homogenous catalysts (those in the same physical state from the reactants) are used in the reaction of a solution of Rochelle salt (potassium sodium tartrate) and hydrogen peroxide. The catalyst is cobalt(II) chloride solution.

The cobalt(II) chloride solution is pink at the start but changes to green as Co^{3+} ions form. Oxygen gas is vigorously given off at this point. At the end of the reaction Co^{2+} ions are regenerated and the pink colour returns.

....

4.7 Summary

Summary

You should now be able to state that:

- atoms and ions of the d block transition metals have an incomplete d subshell of electrons;
- transition metals exhibit variable oxidation states and their chemistry frequently involves redox reactions;
- transition metals form complexes (coordination compounds) which are named according to IUPAC rules;
- the properties of these complexes, such as colour, can be explained by the presence of unfilled and partly filled d orbitals;
- the effects of d $\rightarrow$ d electronic transitions can be studied using ultraviolet and visible absorption spectroscopy, which is an important analytical tool;
- transition metals and their compounds are important as catalysts in many reactions, again due to the presence of a partially filled d subshell.

4.8 Resources

- Royal Society of Chemistry (http://rsc.li/2atncS1)
- Chemguide (http://www.chemguide.co.uk)

4.9 End of topic test

End of Topic 4 test

The end of topic test for *Inorganic and physical chemistry.*

Q36: Which of the following electron configurations could represent a transition metal?

a) $1s^2 2s^2 2p^6 3s^2 3p^6 3d^3 4s^2$
b) $1s^2 2s^2 2p^6 3s^2 3p^5$
c) $1s^2 2s^2 2p^6 3s^2 3p^6 3d^{10} 4s^2 4p^3$
d) $1s^2 2s^2 2p^6 3s^2 3p^6 4s^2$

....

Q37: Part of the electron configuration of iron can be shown thus:

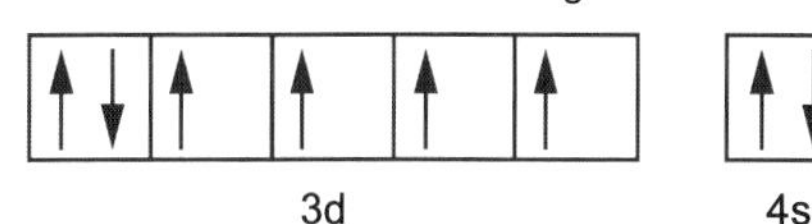

Using the same notation, which of the following shows the correct configuration for a chromium atom?

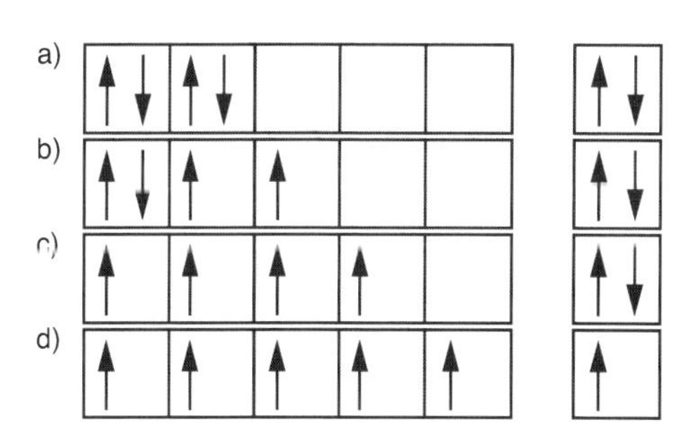

....

Q38: A green hydrated ion has three unpaired electrons. Which of these ions could it be?

a) Fe^{2+}
b) V^{3+}
c) Ni^{2+}
d) Cr^{3+}

....

Q39: What is the oxidation number of nickel in the complex, $Mg_2[NiCl_6]$?

a) +1
b) +2
c) +4
d) +6

....

Q40: Potassium manganate(VII) (potassium permanganate) is purple in colour. In which region of the visible spectrum does it mainly absorb?

a) Red
b) Blue
c) Yellow
d) Purple

....

Q41: $VO^{2+} \rightarrow V^{3+}$

This change involves:

a) an oxidation with loss of one electron.
b) a reduction with gain of one electron.
c) an oxidation with loss of three electrons.
d) a reduction with gain of three electrons.

....

Q42: Which of the following is **true** about ultraviolet spectroscopy?

a) The wavelength range is approximately 400-700 nm.
b) The concentration of the absorbing species can be calculated from the intensity of the absorption.
c) Compounds which absorb only in the ultraviolet are coloured.
d) A UV spectrum is an emission spectrum.

....

The following three questions refer to a complex which has the formula $[Cr(NH_3)_6]Cl_3$.

Q43: Select **two** terms that can be applied to the complex ion.

a) Cation
b) Tetrahedral
c) Anion
d) Octahedral
e) Hexadentate
f) Monodentate

....

Q44: Which term can be applied to the ligand?

a) Cation
b) Tetrahedral
c) Anion
d) Octahedral
e) Hexadentate
f) Monodentate

....

Q45: What is the correct name for this complex?

....

The $[Co(NH_3)_6]^{3+}$ ion is yellow and the $[CoF_6]^{3-}$ ion is blue.

Q46: What is the oxidation state of cobalt in both complex ions?

....

Q47: What is the name of the ligand that causes the stronger crystal field splitting?

....

A solution containing hydrogen peroxide and potassium sodium tartrate was heated. No gas was produced. When pink cobalt(II) chloride was added, the solution turned green and bubbles were produced rapidly. As the bubbling subsided, the green colour turned back to pink.

Q48: What evidence is there to suggest that cobalt(II) chloride acts as a catalyst ?

....

Q49: Which of the following statements explains best how this catalyst works?

a) The catalyst provides a surface on which the reaction takes place.
b) Cobalt forms complexes with different colours.
c) Cobalt exhibits various oxidation states of differing stability.
d) The catalyst provides extra energy.

....

Q50: The $[Co(NH_3)_6]^{3+}$ ion is yellow and the $[CoF_6]^{3-}$ ion is blue.

Explain, possibly with the aid of diagrams, why the ions are different colours.

....

....

Topic 5

Chemical equilibrium

Contents

Prerequisite knowledge

Before you begin this topic, you should be able to:

- *state that the forward and backward reactions in dynamic equilibrium have equal rates and that concentrations of products and reactants will remain constant at this time;*
- *describe how temperature, concentration and pressure affect the position of equilibrium.*

Learning objectives

By the end of this topic, you should be able to:

- *describe the equilibrium chemistry of acids and bases,*
- *write equilibrium expressions;*

- *use the terms: pH, K_w, K_a and pK_a;*
- *understand the chemistry of buffer solutions;*
- *calculate the pH of buffer solutions.*

5.1 Introduction

A chemical reaction is in dynamic equilibrium when the rate of the forward and backward reaction is equal. At this point the concentrations of the reactants and products are constant, but not necessarily equal. From Higher Chemistry you should be aware of the factors that can alter the position of equilibrium including, concentration of reactants or products, pressure and temperature. You should also be aware that a catalyst speeds up the rate at which equilibrium is reached, but does not alter the position of equilibrium.

The nature of chemical equilibrium

A chemical system in equilibrium shows no changes in macroscopic properties, such as overall pressure, total volume and concentration of reactants and products. It appears to be in a completely unchanging state as far as an outside observer is concerned.

Consider a bottle of soda water (carbon dioxide dissolved in water, with free carbon dioxide above). So long as the system remains **closed** , the macroscopic properties (e.g. the pressure of CO_2 in the gas and the concentration of the various dissolved materials) will remain constant - the system is in equilibrium. However, on the microscopic scale there is change. Carbon dioxide molecules in the gas will bombard the liquid surface and dissolve; some carbon dioxide molecules in the solution will have sufficient energy to leave the solution and enter the gas phase.

System in equilibrium

Go online

A system in equilibrium appears to be unchanging as far as an outside observer is concerned. The bottle of soda water or lemonade shown below has carbon dioxide dissolved in the water and also free carbon dioxide above the liquid. The system is **closed** so that nothing can enter or leave the container.

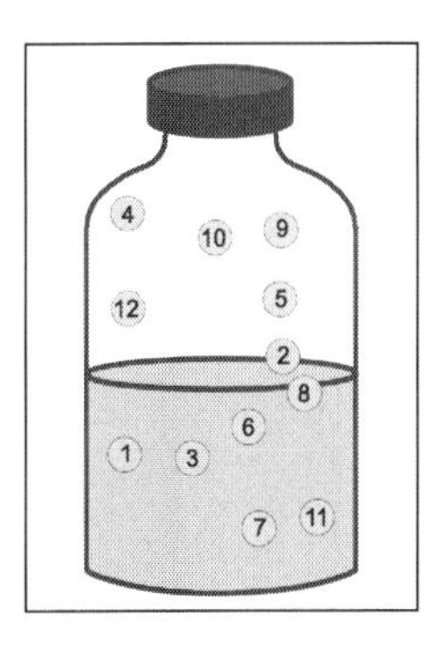

As some carbon dioxide in the gas dissolves, some carbon dioxide in the solution leaves to become gas. So long as the system remains **closed** there is a balance between the rates of the exchange. Notice that the concentrations at equilibrium are not necessarily equal.

....

At equilibrium these two processes will balance and the number of molecules in the gas and liquid will always be the same, although the individual molecules will not remain static. This state is achieved by a **dynamic equilibrium Dynamic equilibrium** between molecules entering and leaving the liquid, and between carbon dioxide, water and carbonic acid. In other words, the rate at which carbonic acid is formed from CO_2 and water will be balanced by carbonic acid dissociating to form CO_2 and water.

$CO_2 + H_2O \rightleftharpoons H_2CO_3$

Hydrogen Iodide equilibrium

Look at the figure below of the reaction between hydrogen and iodine to produce hydrogen iodide in the ***forward reaction***.

$H_2 + I_2 \rightarrow 2HI$

Molecules

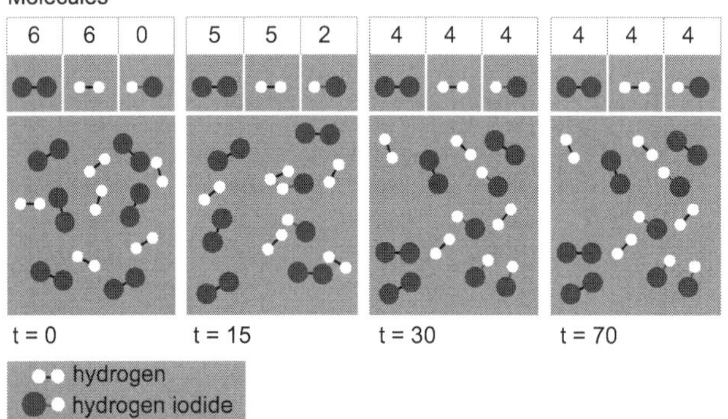

Q1: At t = 0 there are six molecules of H_2, six of I_2 and none of HI. Count the number of molecules of H_2, I_2, and HI after time, t = 15, t = 30 and t = 70. What do you notice about them?

....

After t = 30 there is no further change in the numbers of reactant and product molecules. Therefore this system has reached equilibrium.

The figure below shows only hydrogen iodide molecules of the ***reverse reaction***:

$2HI \rightarrow H_2 + I_2$

Molecules

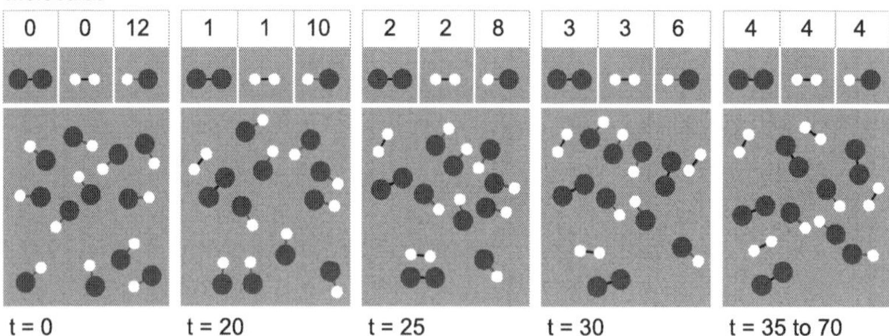

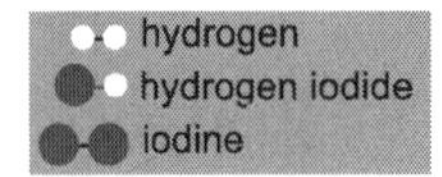

In this case, two molecules of hydrogen iodide react to form hydrogen and iodine.

Q2: What do you notice about the number of H_I, H_2 and I_2 molecules after time, t = 35 and t = 70?

....

Q3: But how do these numbers at equilibrium compare with the previous reaction starting with hydrogen and iodine?

....

....

At equilibrium, the rate of production of HI from H_2 and I_2 equals the rate of production of H_2 and I_2 from HI, therefore the overall composition will not change. This process is generally shown by the use of reversible arrows.

$H_2 + I_2 \rightleftharpoons 2HI$

The graph shows the progress of the reaction starting with H_2 + I_2 to produce an equilibrium mixture of reactants and products.

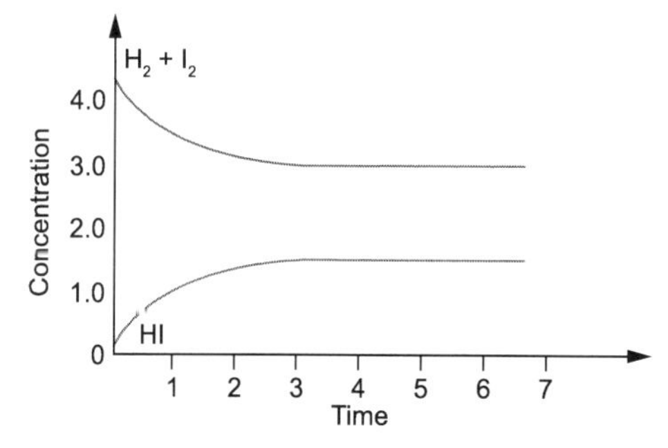

The next graph shows the progress of the decomposition of HI under the same conditions.

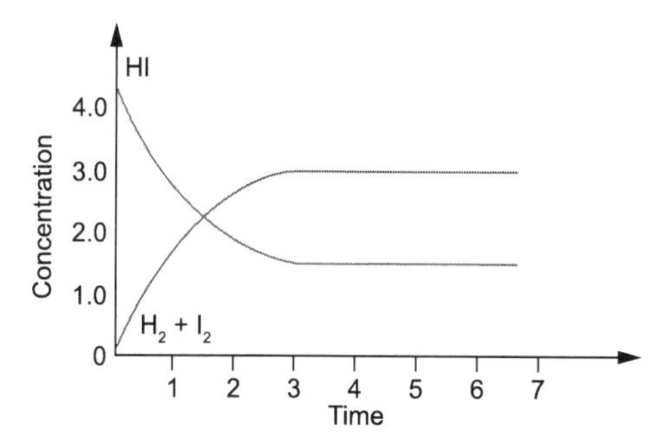

You will notice that the same equilibrium mixture is obtained, whether you start with hydrogen and iodine or with hydrogen iodide.

Key point

- A chemical reaction is at equilibrium when the composition of the reactants and products remains constant indefinitely.
- This state occurs when the rates of the forward and reverse reactions are equal.
- The same equilibrium mixture is obtained whether you start with reactants or products.

Consider a simple case of physical equilibrium. A boulder at the top of a hill will remain there until disturbed - it is in a state of equilibrium. When pushed, however, it will readily move into the valley where it will remain even when displaced slightly. The boulder's unstable equilibrium position, shown in the following figure, has moved to a stable equilibrium. This state usually represents a minimum energy state (in this case, the lowest gravitational potential energy).

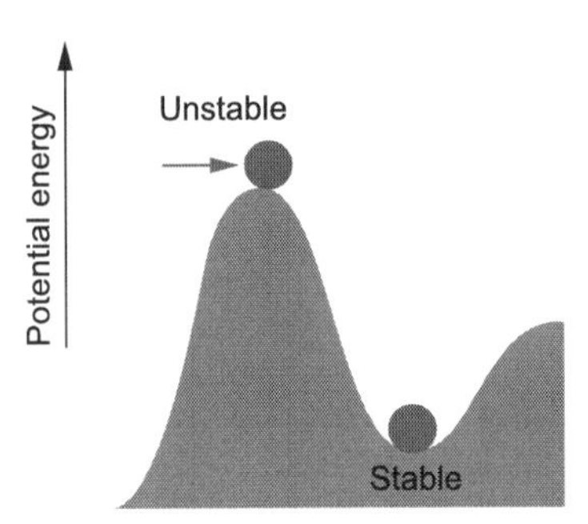

5.2 Equilibrium expressions and factors affecting equilibrium

The equilibrium constant is given the symbol K. It is written as Kc when describing the equilibrium in terms of concentration.

aA + bB $\rightleftharpoons$ cC + dD

In the equation

$$Kc = \frac{[C]^c[D]^d}{[A]^a[B]^b}$$

where [A], [B], [C] and [D] are the equilibrium constants of A, B, C and D respectively and a, b, c and d are the stoichiometric coefficients in a balanced chemical reaction.

The balanced equation for the Haber process is $N_2(g) + 3H_2(g) \rightleftharpoons 2NH_3(g)$

and therefore the equilibrium constant expression is

$$Kc = \frac{[NH_3]^2}{[N_2]\,[H_2]^3}$$

Concentration values are usually measured in mol l^{-1}.

Since the equilibrium constant is the ratio of concentration of products divided by the concentration of reactants, its actual value gives guidance to the extent of a reaction once it has reached equilibrium. The greater the value of Kc the greater the concentration of products compared to reactants; in other words, the further the reaction has gone to completion.

The explosive reaction between hydrogen and fluorine:

$H_2 + F_2 \rightleftharpoons 2HF$

has an equilibrium constant of 1×10^{47}. At equilibrium, negligible amounts of the reactants will remain; almost all will have been converted to hydrogen fluoride.

In contrast, the dissociation of chlorine molecules to atoms:

$Cl_2 \rightleftharpoons 2Cl$

has a K_c value of 1×10^{-38} at normal temperatures, indicating a reaction which hardly occurs at all under these circumstances.

In time, all reactions can be considered to reach equilibrium. To simplify matters, the following general assumption is made:

Value of Kc	Extent of reaction
$< 10^{-3}$	Effectively no reaction
10^{-3} to 10^{3}	Significant quantities of reactants and products at equilibrium
$> 10^{3}$	Reaction is effectively complete

A note of caution:

The equilibrium constant gives no indication of the rate at which equilibrium is achieved. It indicates only the ratios of products to reactants once this state is reached.

Q4: Write an equilibrium expression for the following reactions.
$2Fe^{3+}(aq) + 3I^-(aq) \rightleftharpoons 2Fe^{2+}(aq) + I_3^-(aq)$

....

Q5: Write an equilibrium expression for the following reactions.
$H_3PO_4(aq) \rightleftharpoons 2H^+(aq) + HPO_4^{2-}(aq)$

....

Returning to the reaction:

$$H_2 + I_2 \rightleftharpoons 2HI$$

the equilibrium constant is defined as:

$$Kc = \frac{[HI]^2}{[H_2][I_2]}$$

and at 453°C, it has a value of 50.

Q6: At 453°C which compound is present in greatest concentration?

a) Hydrogen
b) Iodine
c) Hydrogen iodide
d) All the same concentration

....

Q7: The Kc value for the reaction
$PCl_5 \rightleftharpoons PCl_3 + Cl_2$
is 0.021 at 160°C. Which compound is present in greatest concentration at equilibrium?

a) Phosphorus(V) chloride
b) Phosphorus(III) chloride
c) Chlorine
d) All are the same

....

Q8: The following equilibrium constants apply at room temperature (25°C).
$Zn(s) + Cu^{2+}(aq) \rightleftharpoons Cu(s) + Zn^{2+}(aq)$ $K = 2 \times 10^{37}$
$Mg(s) + Cu^{2+}(aq) \rightleftharpoons Cu(s) + Mg^{2+}(aq)$ $K = 6 \times 10^{90}$
$Fe(s) + Cu^{2+}(aq) \rightleftharpoons Cu(s) + Fe^{2+}(aq)$ $K = 3 \times 10^{26}$
Of the metals Zn, Mg, and Fe, which removes Cu(II) ions from solution most completely?

a) Zn
b) Mg
c) Fe

....

Q9: In which of the following reactions will the equilibrium lie furthest towards products?

a) $N_2O_4(g) \rightleftharpoons 2NO_2(g)$ K_c at 0°C = 159
b) $2SO_2(g) + O_2(g) \rightleftharpoons 2SO_3(g)$ K_c at 856°C = 21.1
c) $N_2O_4(g) \rightleftharpoons 2NO_2(g)$ K_c at 25°C = 14.4
d) $2SO_2(g) + O_2(g) \rightleftharpoons 2SO_3(g)$ K_c at 636°C = 3343

....

Q10: From the data in the previous question, what do you notice about the value of K_c for the oxidation of SO_2 at 856°C compared with 636°C?

....

Q11: Would the process to manufacture SO_3 be more productive at:

a) 636°C
b) 856°C

....

For gaseous reactions partial pressures may be used. Gases inside a closed container each exert a pressure proportional to the number of moles of the particular gas present (for example, if two gases are mixed in equimolar amounts and the total pressure is 1 atmosphere, then the partial pressure of each gas is 0.5 atmosphere).

$$N_2(g) + 3H_2(g) \rightleftharpoons 2NH_3(g)$$

$$Kp = \frac{(pNH_3)^2}{(pN_2)(pH_2)^3}$$

Q12: Write down an appropriate expression for the equilibrium constant for the following reactions:
$2NOCl(g) \rightleftharpoons 2NO(g) + Cl_2(g)$

....

Q13: Write down an appropriate expression for the equilibrium constant for the following reactions:
$2SO_2(g) + O_2(g) \rightleftharpoons 2SO_3(g)$

....

The equilibrium constant has no units whatever the concentrations are measured in.

The above example of the Haber process is an example of a homogenous equilibrium where all the species are in the same phase. In heterogeneous equilibria the species are in different phases; an example showing this is through heating calcium carbonate in a closed system. The carbon dioxide formed cannot escape, setting up equilibrium.

$$CaCO_3(s) \rightleftharpoons CO_2(g) + CaO(s)$$

In this reaction where solids exist at equilibrium their concentration is taken as being constant and given the value of 1.

So instead of

$$Kc = \frac{[CO_2(g)][CaO(s)]}{[CaCO_3(s)]}$$

the equilibrium expression is written as K = [CO_2(g)].

This is also true for pure liquids (including water) where their equilibrium concentration is given the value of 1. This is not true however for aqueous solutions.

Changing the concentration or pressure has an effect on the position of equilibrium (Higher Chemistry Unit 3) however, the equilibrium constant K is not affected.

For example NH_3(g) + H_2O(l) $\rightleftharpoons$ NH_4^+(aq) + OH^-(aq)

$$K = \frac{[NH_4{}^+][OH^-]}{[NH_3]}$$

If more ammonium ions in the form of solid ammonium chloride are added to the equilibrium the position will shift to the left since ammonium ions are present on the right hand side of the equilibrium (follows Le Chatelier's principle). This increases the concentration of the ammonium ions which causes the system to react in order to decrease the concentration to restore equilibrium. This alters the position of equilibrium until the ratio of products to reactants is the same as before re-establishing the value of K.

Changes in pressure only affect reactions involving gases. If the pressure is increased then the position of equilibrium will shift to the side with the fewer number of gaseous moles. This then causes a new equilibrium to be established, but with the same value of K.

Changes in temperature affect the value of the equilibrium constant K as it is temperature dependent.

Reactants $\rightleftharpoons$ Products

If the forward reaction is exothermic an increase in temperature favours an increase in the concentration of the reactants affecting the ratio [products]/[reactants] by decreasing it. The value of K therefore is decreased.

If the forward reaction is endothermic an increase in temperature favours an increase in the concentration of the products increasing the ratio [products]/[reactants] and increasing the value of K.

5.3 Phase equilibria

Partition coefficients

Immiscible liquids do not mix with each other, the liquid with the lesser density floating on the liquid with the greater density.

Two immiscible liquids are cyclohexane and an aqueous solution of potassium iodide. Solid iodine dissolves in both these liquids and when shaken in both these liquids some dissolves in one liquid while the remainder stays dissolved in the other. The iodine partitions itself between the two liquids. Some of the solute dissolved in the lower level starts to move into the upper layer while at the same time solute in the upper layer starts to move to the lower layer. Eventually the rate of movement from the lower layer to the upper level becomes the same as the rate of the movement from the upper layer to the lower layer and a dynamic equilibrium is set up.

$$I_2(aq) \rightleftharpoons I_2(C_6H_{12})$$

$$K = \frac{[I_2(C_6H_{12})]}{[I_2(aq)]}$$

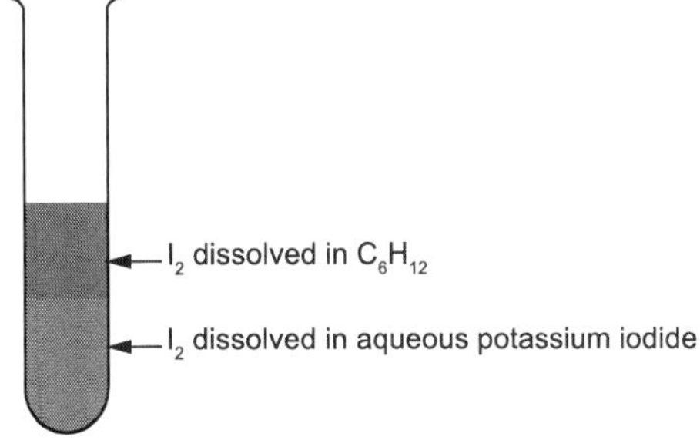

Equilibrium constant is known as the partition coefficient and is temperature dependent. It is affected by what solvents/solute are used, but not by adding more solvent or solute.

Solvent extraction

Partition can be used to extract and purify a desired product from a reaction mixture using a separating funnel. This method relies on the product being more soluble in one liquid phase than the other. Caffeine for example is more soluble in dichloromethane than water and this can be used to produce decaffeinated coffee. Duo to thc harmful nature of dichloromethane caffeine is now extracted using supercritical carbon dioxide, which acts like both a liquid and a gas. It is more efficient to use smaller quantities of the liquid carrying out the extraction a few times rather than using the whole volume at once.

Suppose that for a carboxylic acid partitioning between ether and water:

$$K = \frac{[acid] in ether}{[acid] in water} = \frac{[Ae]}{[Aw]} = 5$$

If 10 g of acid is dissolved in 100 cm^3 of water and 100 cm^3 of ether is available for extraction, the difference in the quantity extracted by 100 cm^3 of ether used in one extraction of 100 cm^3 or in four extractions of 25 cm^3 can be calculated as follows.

Let *v* g of acid be extracted with 100 cm^3 ether in a single extraction:

$$K = \frac{[Ae]}{[Aw]} = 5$$

$$\frac{v/100}{10-v/100} = 5$$

$$\frac{v}{10-v} = 5$$

v = 8.33 g, i.e. 8.33 g is extracted into the ether

When 4 $\times$ 25 cm^3 portions of ether are used, the calculation has to be repeated four times.

Let *w, x, y* and *z* g of acid be extracted in each successive extraction.

First extraction

$$K = \frac{[Ae]}{[Aw]} = 5$$
$$\frac{w/25}{10-w/100} = 5$$
$$4w = 50 - 5w$$
$$w = 5.56g$$

This means that 4.44 g remains in the water to be extracted by the next 25 cm^3 of ether.

Second extraction

$$\frac{x/25}{4.44-x/100} = 5$$
$$4x = 22.2 - 5x$$
$$x = 2.47g$$

Now 1.97 g remains.

Third extraction

$$\frac{y/25}{1.97-y/100} = 5$$
$$4y = 9.85 - 5y$$
$$y = 1.09g$$

Now 0.88 g remains.

Fourth extraction

$$\frac{z/25}{0.88-z/100} = 5$$
$$4z = 4.4 - 5z$$
$$z = 0.49g$$

Total amount of carboxylic acid extracted

= w + x + y + z
= 5.56 + 2.47 + 1.09 + 0.49
= 9.61 g

This calculation shows that an extra 1.28 g (9.61 - 8.33 g) of carboxylic acid can be extracted when 4 $\times$ 25 cm^3 extractions are used rather than 1 $\times$ 100 cm^3 extraction.

5.3.1 Chromatography

All chromatographic methods involve a mobile phase moving over a stationary phase. Separation occurs because the substances in the mixture have different partition coefficients between the stationary and mobile phases.

Substances present in the initial mixture which partition more strongly into the stationary phase will move more slowly than materials which partition more strongly into the mobile phase.

In paper chromatography the mixture of components to be separated is placed as a small spot close to the bottom of a rectangular piece of absorbent paper (like filter paper). The bottom of the paper is placed in a shallow pool of solvent in a tank. An example of the solvent would be an alcohol. Owing to capillary attraction, the solvent is drawn up the paper, becoming the mobile phase. The solvent front is clearly visible as the chromatography progresses.

When the paper is removed from the solvent, the various components in the initial spot have moved different distances up the paper.

Paper chromatography

A simple simulation of this process, showing chromatography of blue and black inks, is available on the on-line version of this Topic. The start and final positions are shown in Figure 5.1 and Figure 5.2 respectively.

Figure 5.1: Start of chromatographic analysis

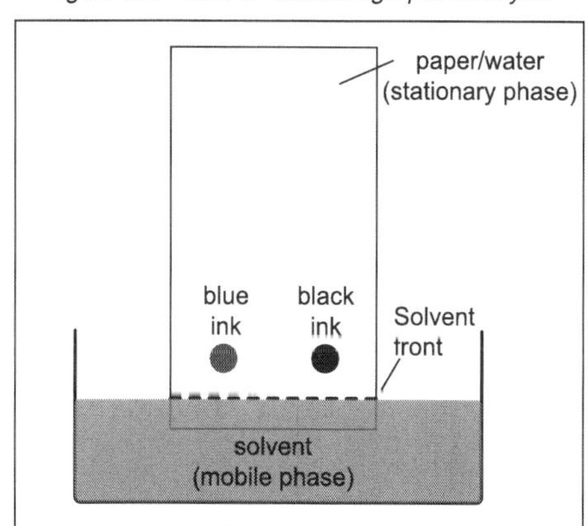

....

Figure 5.2: End of chromatographic analysis

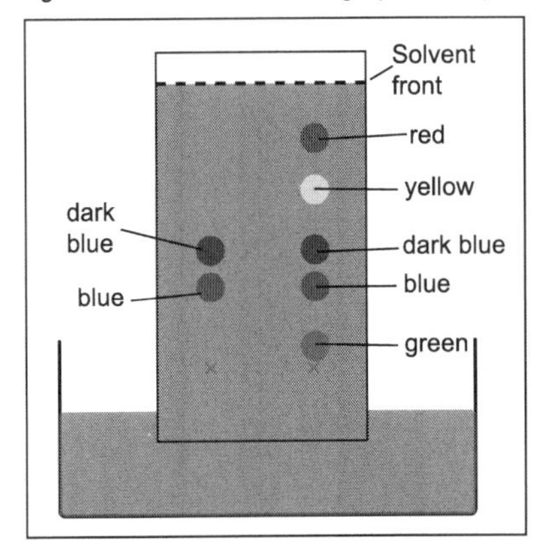

....

Q14: By observing the chromatography simulation above, which ink (blue or black) has the most components?

a) Blue
b) Black
c) Both the same

....

Q15: Which material in the black ink has stayed longest on the stationary phase, and has the lowest solvent/water partition coefficient?

a) Red
b) Yellow
c) Dark Blue
d) Blue
e) Green

....

Q16: The red coloured spot has moved furthest, this would indicate that the red material:

a) has the highest solvent/water partition coefficient.
b) has the lowest solvent/water partition coefficient.
c) has the lowest molecular mass.

....

Q17: The movement of materials on paper chromatography is often described by an R_f value which is the distance travelled by the spot divided by the distance travelled by the solvent front. As long as the conditions of chromatography remain the same, a compound will have a constant R_f value.

Which colour in the black ink could have an R_f value of 0.4?

a) Red
b) Yellow
c) Dark blue
d) Blue
e) Green

....

Q18: Both the blue and dark blue spots from both the original inks have moved similar distances. What might you conclude from this?

....

....

This separation depends on the different partition coefficients of the various components. The components are partitioned between the solvent and the water trapped in the paper. Substances which partition mainly into the solvent mobile phase will move further up the paper than substances which partition more strongly onto the stationary phase.

Although paper chromatography is still used today, it has been largely replaced by thin layer chromatography (TLC). In this method, a support of glass or aluminium is coated, usually with a thin layer of silica or cellulose. The processing is identical to that described above, but TLC allows a more rapid separation (which prevents the spots spreading too far) and makes detection of the spots easier. Most materials are not coloured, but can still be chromatographed.

The invisible spots on paper or thin layer chromatography are revealed by use of a locating reagent. These react with the compounds in the spots to produce a coloured derivative. For example, ninhydrin solution can be sprayed onto chromatograms to reveal amino-acids.

In another TLC detection system, the silica stationary phase is mixed with a fluorescent dye, so that at the end of the process, viewing the plate under an ultraviolet lamp will cause the background to glow (often an eerie green) **except** where there are spots, which appear black.

In gas-liquid chromatography (**GLC**) the stationary phase is a high boiling point liquid held on an inert, finely-powdered support material, and the mobile phase is a gas (often called the carrier gas). The stationary phase is packed into a tubular column usually of glass or metal, with a length of 1 to 3 metres and internal diameter about 2 mm. One end of the column is connected to a gas supply (often nitrogen or helium) via a device which enables a small volume of liquid sample (containing the mixture to be analysed) to be injected into the gas stream. The other end of the column is connected to a device which can detect the presence of compounds in the gas stream. The column is housed in an oven to enable the temperature to be controlled throughout the chromatographic analysis. One reason for this is that the materials to be analysed must be gases during the analysis, so that gas-liquid chromatography is often carried out at elevated temperatures.

A mixture of the material to be analysed is injected into the gas stream at zero time. The individual components travel through the packed column at rates which depend on their partition coefficients between the liquid stationary phase and the gaseous mobile phase. The detector is set to measure some change in the carrier gas that signals the presence of material coming from the end of the column. Some detectors measure the thermal conductivity of the gas, others burn the material from the column in a hydrogen-air flame and measure the presence of ions in the flame. The signal from the detector is recorded and plotted against time to give a series of peaks each with an individual retention time.

Figure 5.3: GLC apparatus

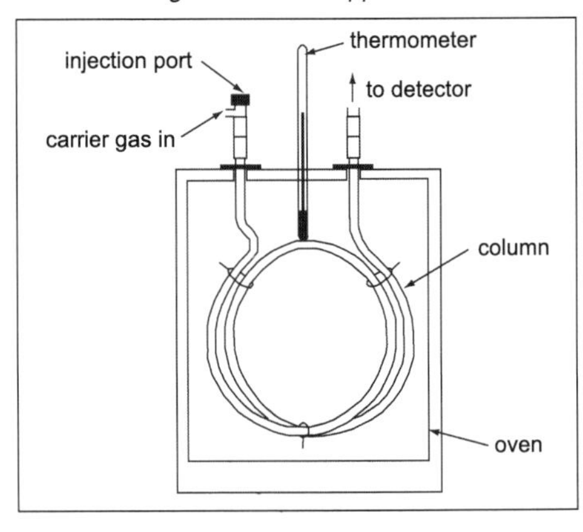

. . . .

5.4 Acid/base equilibria

In acids the concentration of hydrogen ions (H^+) is greater than the concentration of hydrogen ions in water. The concentration of hydroxide ions (OH^-) is greater in alkalis than the concentration of hydroxide ions in water.

A hydrogen ion is basically a proton (hydrogen atom that has lost an electron) and only exists when surrounded by water molecules in an aqueous solution. They are written as $H_3O^+(aq)$ but are often shortened to just $H^+(aq)$.

An acid therefore is a proton donor and a base is a proton acceptor. This definition was put forward by Bronsted and Lowry in 1932. When an acid donates a proton the species left is called the **conjugate base**.

$HA \rightleftharpoons$	H^+ +	A^-
Acid		Conjugate base

When a base accepts a proton the species formed is a **conjugate acid** .

$B + H^+ \rightleftharpoons$ BH^+

Base Conjugate acid

Acid	**Base**	**Conjugate acid**	**Conjugate base**
HNO_3	H_2O	H_3O^+	NO_3^-
HCOOH	H_2O	H_3O^+	$HCOO^-$
H_2O	NH_3	NH_4^+	OH^-
H_2O	F^-	HF	OH^-

5.4.1 Ionisation of water

From the above table we can see that water can act as both an acid and a base. Therefore water can be called **amphoteric** i.e. can behave as either a base or an acid.

In water and aqueous solutions there is an equilibrium between the water molecules, and hydrogen and hydroxide ions. Water acts both as a proton donor (acid) forming a conjugate base and the proton acceptor (base) forming a conjugate acid.

This (dissociation) ionisation of water can be represented by:

$H_2O(\ell)$ (acid) + $H_2O(\ell)$ (base) $\rightleftharpoons$ $H_3O^+(aq)$ + $OH^-(aq)$

Conjugate acid Conjugate base

The equilibrium constant K is defined by K = $[H_3O^+(aq)]$ $[OH^-(aq)]$ or simply $[H^+][OH^-]$ and is known as the **ionic product of water** and given the symbol Kw. This is temperature dependent and the value is approximately 1×10^{-14} at 25°C.

5.4.2 pH scale

Key point

pH values are not always whole numbers.

The pH scale

The concentration of H^+ and OH^- in pure water is very small. At 25°C the concentration of H^+ (and OH^-) is 0.0000001 in units of mole per litre. The square brackets around the symbol H^+ i.e. $[H^+]$ is the *concentration in mole per litre*. In scientific notation this is 1.0×10^{-7} mol ℓ^{-1}.

Rather than use these very small fractions, chemists convert the H^+ concentrations to a new scale - the pH scale- which uses small positive numbers. The way to do this is to define the pH as:

$$pH = -\log_{10}[H^+]$$

H^+ = Hydrogen ion concentrat

$[H^+]$	10^{-2}	10^{-4}	10^{-6}	**10^{-7}**	10^{-8}	10^{-10}	10^{-12}	10^{-14}
$\log_{10}[H^+]$	-2	-4	-6	**-7**	-8	-10	-12	-14
$-\log_{10}[H^+]$	+2	+4	+6	**+7**	+8	+10	+12	+14
pH	2	4	6	**7**	8	10	12	14

Acids pH values lower than 7 $[H^+] > [OH^-]$

Bases pH values higher than 7 $[H^+] < [OH^-]$

Neutral solutions pH values = 7 $[H^+] = [OH^-]$

The relationship between pH and hydrogen ion concentration is given by the equation:

$pH = -\log_{10}[H^+]$

This can be used to calculate the pH of strong acids and alkalis (see later for information on these).

Examples

1. Calculate the pH

Calculate the pH of 0.21mol l^{-1} HCl (aq)

$[H^+] = 0.21 mol\ l^{-1}$ $pH = -\log_{10}(0.21) = 0.68$

You should also be able to calculate concentrations from the pH value.

....

2. [H+] from a pH value

Calculate $[H^+]$ from a pH value of 11.6

$11.6 = -\log_{10}[H^+]$ ($[H^+] = 10^{-pH}$ for students who have difficulty converting between logs and powers of 10)

$[H^+] = 2.51 \times 10^{-12}$ mol l^{-1}.

The pH of neutral solutions can be calculated to be 7 as shown below:

$Kw = [H^+][OH^-] = 1x10^{-14}$

As in neutral solutions $[H^+] = [OH^-]$ then both the concentration of H^+ and OH^- are $1x10^{-7}$.

$pH = -\log_{10}(1x10^{-7}) = 7$.

This pH of 7 for neutral solutions only applies at 25°C. The pH will decrease as temperature increases due to the fact that the ionisation of water is an endothermic process, therefore the value of Kw will increase with increasing temperature. See equilibrium constants changing with temperature in the section 'Equilibrium expressions and factors affecting equilibrium' earlier in this topic.

....

Calculating pH

Go online

Key point

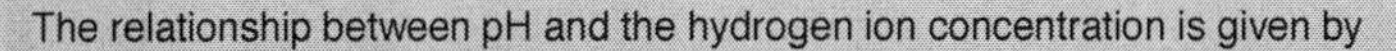

The relationship between pH and the hydrogen ion concentration is given by

$$pH = -\log_{10}[H^+]$$

This relationship can be used to calculate the pH for strong acids and alkalis given the molar concentrations of either H^+ or OH^- ions.

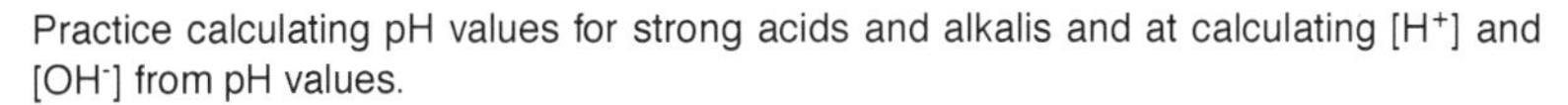

Practice calculating pH values for strong acids and alkalis and at calculating $[H^+]$ and $[OH^-]$ from pH values.

Examples

1. pH from [H⁺] (1)

Calculate the pH of a 0.02 mol ℓ^{-1} solution of hydrochloric acid.

$$[H^+] = \quad 0.02moll^{-1}$$
$$= 2x10^{-2}moll^{-1}$$
$$-\log[H^+] = 1.7$$
$$pH = 1.7$$

It is always useful to check your answer by estimating values.

0.02 lies between 0.01 and 0.10, i.e. between 10^{-2} and 10^{-1}.

So the pH must lie between 2 and 1.

....

2. pH from [H⁺] (2)

Calculate the pH of a 0.001 mol ℓ^{-1} solution of sulfuric acid.

Since there are 2 moles of H^+ ions present per mole of H_2SO_4 then:

$$[H^+] = \quad 0.002moll^{-1}$$
$$= 2x10^{-3}moll^{-1}$$
$$-\log[H^+] = 2.7$$
$$pH = 2.7$$

....

3. pH from [OH⁻]

Calculate the pH of a solution of 0.006 mol ℓ^{-1} sodium hydroxide.

$$
\begin{aligned}
[OH^-] &= 0.006 \\
&= 6 \times 10^{-3} moll^{-1} \\
K_w &= [H^+]\,[OH^-] \\
[H^+] &= \frac{K_w}{[OH^-]} \\
&= \frac{10^{-14}}{6\times10^{-3}} \\
&= 1.667 \times 10^{-12}
\end{aligned}
$$

$$
\begin{aligned}
pH &= -\log 1.667 - \log\left(10^{-12}\right) \\
&= -0.22 + 12 \\
&= \underline{\underline{11.78}}
\end{aligned}
$$

Check [H^+] lies between 10^{-11} and 10^{-12}, i.e pH lies between 11 and 12.

. . . .

Q19: Calculate the pH of a solution that has a H^+(aq) concentration of 5 $\times$ 10^{-3} mol ℓ^{-1}. Give your answer to 2 decimal places.

. . . .

Q20: Calculate the pH of a solution that has a H^+(aq) concentration of 8 $\times$ 10^{-6} mol ℓ^{-1}. Give your answer to 2 decimal places.

. . . .

Q21: Calculate the pH of a solution that has a OH^-(aq) concentration of 6.3 $\times$ 10^{-2} mol ℓ^{-1}. Give your answer to 2 decimal places.

. . . .

Q22: Calculate the pH of a solution that has a OH^-(aq) concentration of 2.9 $\times$ 10^{-4} mol ℓ^{-1}. Give your answer to 2 decimal places.

. . . .

Using the same relationship ($pH = -\log_{10}[H^+]$), the concentration of H^+ ions and OH^- ions can be calculated from the pH of the solution.

Example : Concentrations from pH

Calculate the concentration of H^+ ions and OH^- ions in a solution of pH 3.6. Give your answer to three significant figures.

Step 1

$$\begin{aligned} pH = & \quad -\log[H^+] \\ 3.6 = & \quad -\log[H^+] \\ \log[H^+] = & \quad -3.6 \\ [H^+] = & \quad \text{antilog}(-3.6) \\ = & \quad 10^{-3.6} \\ = & \quad 0.000251 \\ = & \quad \underline{\underline{2.51 \times 10^{-4}}} moll^{-1} \end{aligned}$$

Note: calculators vary slightly in the way in which they antilog numbers. One way is as follows:

$$\begin{aligned} \text{If} \log[H^+] = & \quad x \\ \text{then} [H^+] = & \quad 10^x \end{aligned}$$

Press the 10^x key, then type the value (-3.6), followed by '='.

If this does not work with your calculator, see your tutor.

Step 2

$$\begin{aligned} K_w = & \quad [H^+][OH^-] \\ [OH^-] = & \quad \frac{K_w}{[H^+]} \\ = & \quad \frac{10^{-14}}{2.51\text{x}10^{-4}} \\ = & \quad \underline{3.98\text{x}10^{-11}} moll^{-1} \end{aligned}$$

....

An alternative method uses the equation pH + pOH = 14.

Step 1 is exactly the same using this method.

Step 2:

pH + pOH = 14
pOH = 14 - 3.6 = 10.4
-log[OH^-] = 10.4
[OH^-] = $10^{-10.4}$
= 3.98×10^{-11}

Q23: Calculate the concentration of H^+(aq) ions and OH^-(aq) ions in a solution of pH 2.3

....

Q24: Calculate the concentration of H^+(aq) ions and OH^-(aq) ions in a solution of pH 5.6

....

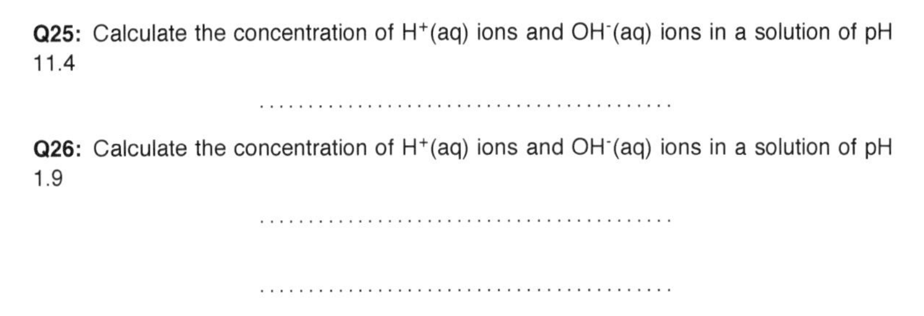

Q25: Calculate the concentration of H^+(aq) ions and OH^-(aq) ions in a solution of pH 11.4

....

Q26: Calculate the concentration of H^+(aq) ions and OH^-(aq) ions in a solution of pH 1.9

....

....

5.4.3 Strong/weak acids and bases

Strong acids and bases are those that completely ionise (dissociate) when dissolved in water. Examples of strong acids include hydrochloric acid, sulfuric acid and nitric acid.

$$HCl(g) \rightarrow H^+(aq) + Cl^-(aq)$$

Hydrochloric acid completely dissociates into hydrogen ions and chloride ions when dissolved in water.

Examples of strong bases include sodium hydroxide, potassium hydroxide and calcium hydroxide.

$$NaOH(s) \rightarrow Na^+(aq) + OH^-(aq)$$

Sodium hydroxide completely dissociates into sodium ions and hydroxide ions when dissolved in water.

For strong monoprotic acids (acids with only one hydrogen ion in their formula, e.g. HCl), the hydrogen ion concentration is the same as the original concentration of the acid as all the acid molecules have dissociated into ions. For strong diprotic acids (acids with two hydrogen ions in their formula, e.g. H_2SO_4), the hydrogen ion concentration will be double that of the original acid concentration since all the hydrogen ions will be released in solution.

Weak acids are only partially ionised (dissociated) when dissolved in water and an equilibrium is set up which lies to the left. Approximately only 1% of the acid molecules are dissociated and therefore the hydrogen ion concentration in solution will be much less than the concentration of the acid. Examples of weak acids include ethanoic acid (all carboxylic acids), carbonic acid and sulfurous acid.

$$CH_3COOH(l) + H_2O(l) \rightleftharpoons CH_3COO^-(aq) + H_3O^+(aq)$$

This is also true for weak bases. Examples of weak bases include ammonia (NH_3) and amines (CH_3NH_2).

$$NH_3(g) + H_2O(l) \rightleftharpoons NH_4^+(aq) + OH^-(aq)$$

We can compare conductivity, pH, rate of reaction and volume to neutralise alkali of strong and weak acids.

Property	Strong acid	Weak acid
Conductivity	Higher	Lower
Rate of Reaction	Faster	Slower
pH	Lower	Higher
Volume to neutralise acid	Same	Same

The differences in conductivity, pH and rate of reaction can be attributed to the fact that strong acids have a much higher number of hydrogen ions in solution than weak acids of the same concentration. However, the volume of alkali required to neutralise a strong and a weak acid of the same concentration is the same. The hydroxide ions in the alkali react with all of the available hydrogen ions in solution. However, in a weak acid, this removes hydrogen ions from the equilibrium and causes the acid molecules to release more hydrogen ions. This continues until all the acid molecules have dissociated, i.e. until the acid is neutralised. The volume of alkali required therefore depends only on the concentration of the acid and not on the strength of the acid.

We can compare conductivity, pH and volume to neutralise acid of weak and strong bases.

Property	Strong base	Weak base
Conductivty	Higher	Lower
pH	Higher	Lower
Volume to neutralise alkali	Same	Same

To calculate the pH of weak acids we need to use the formula pH = $\frac{1}{2}$ pKa - $\frac{1}{2}$ log c

$$HA \rightleftharpoons H^+ + A^-$$

At equilibrium the $[H^+] = [A^-]$ and since the equilibrium lies very far to the left hand side, [HA] at equilibrium is approximately the same as the original concentration of the acid, c. This can therefore be written as $Ka = [H^+]^2/c$.

Therefore $[H^+]^2 = Kac$ and $[H^+] = \sqrt{Kac}$.

Since pH = -log$[H^+]$, it follows that pH = -log $\sqrt{Kac}$.
As -logKa = pKa, pH = -$\frac{1}{2}$logKa - $\frac{1}{2}$logc.
The equation then becomes pH = $\frac{1}{2}$pKa - $\frac{1}{2}$logc.

The larger the value of Ka the stronger the acid (or vice versa) and the smaller the pKa the stronger the acid.

If we use the above equation as a general formula for a weak acid dissociating we can write the equilibrium constant of the acid as

$$Ka = \frac{[H^+][A^-]}{[HA]}$$

$$[H^+] = \sqrt{Kac}$$

where c is the concentration of the acid.

The dissociation constant of a weak acid can be represented by pKa = -logKa which is often more convenient to use than Ka and relates to the equation above for pH of a weak acid.

In weak bases the dissociation constant is written as:

$$Kb = \frac{[NH_4{}^{+}]\,[OH^{-}]}{[NH_3]}$$

The ammonium ion formed is a weak acid and will dissociate as follows:

$$NH_4{}^{+}(aq) + H_2O(l) \rightleftharpoons NH_3(aq) + H_3O^{+}(aq)$$

This time the H_2O is the base, conjugate base is NH_3 and the conjugate acid H_3O^+.

Dissociation constant Ka for the ammonium ion is

$$Ka = \frac{[NH_3]\,[H_3O^{+}]}{[NH_4{}^{+}]}$$

The greater the numerical value of Ka for a weak acid the stronger it is. The smaller the pKa value, the stronger the acid. Ka and pKa values are given in page 13 of the Chemistry Data Booklet (http://bit.ly/29TG6Wr).

Q27: Calculate the pH of a 0.1 mol ℓ^{-1} solution of ethanoic acid. Answer to two decimal places.

....

5.4.4 Salts

Salts are most simply defined as one of the products of the neutralisation of an acid by a base. More accurately, a salt is formed when the hydrogen ions of an acid are replaced by metal ions or ammonium ions.

The first part of the name of a salt identifies which alkali/base was used to make the salt. Sodium salts are generally made using sodium hydroxide and potassium salts made using potassium hydroxide as the alkali.

The second part of the name of a salt identifies the acid used to make the salt. Chloride salts are made from hydrochloric acid, nitrate salts nitric acid and sulfate salts sulfuric acid.

Q28: Identify the parent acid and base used to form magnesium nitrate.

....

Q29: Identify the parent acid and base used to form potassium bromide.

....

Q30: Identify the parent acid and base used to form sodium ethanoate.

....

Q31: Identify the parent acid and base used to form calcium sulfite. (Care - note the different name ending.)

....

Some salt solutions are neutral but not all. The pH of a salt solution depends on the relative strengths of the parent acid and parent base. You can imagine the acid trying to pull the pH towards the acidic side but being opposed by the base pulling in the opposite direction, as in a tug of war. Whichever is stronger will pull the pH towards its end of the scale.

Figure 5.4: pH scale and universal indicator colour

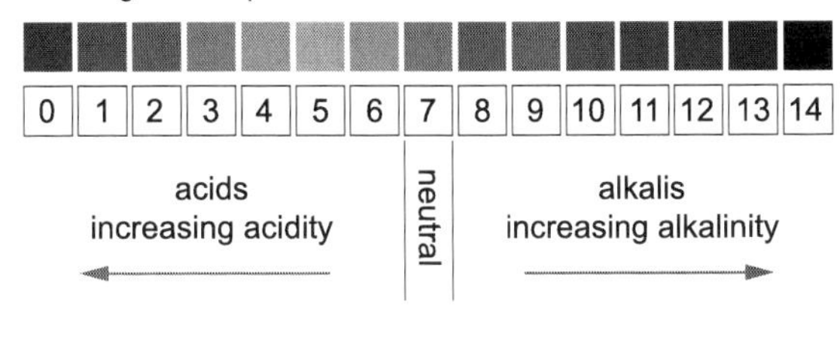

....

- If the parent acid is stronger than the parent base, the pH of the salt solution will be less than 7, i.e. acidic.
- If the parent base is stronger than the parent acid, the pH of the salt solution will be greater than 7, i.e. alkaline.
- If the acid and base are equally strong, the salt solution will be neutral (pH 7).

For example potassium ethanoate will have been made using the parent acid ethanoic acid (weak acid) and the alkali potassium hydroxide (strong base) and therefore the pH of potassium ethanoate will be more than 7 and alkaline.

Q32: Which of the following is likely to be the pH of a solution of sodium ethanoate?

a) 5
b) 7
c) 9

....

Q33: Soaps are salts of fatty acids and strong bases such as sodium hydroxide or potassium hydroxide. Will a soap solution be acidic, neutral or alkaline?

....

Q34: A solution of the salt, potassium cyanide (KCN), has a pH of between 8 and 9. Which of the following is true?

a) The acid, hydrogen cyanide, is strong and the base is weak.
b) The acid, hydrogen cyanide, is strong and the base is strong.
c) The acid, hydrogen cyanide, is weak and the base is weak.
d) The acid, hydrogen cyanide, is weak and the base is strong.

....

Q35: Predict the pH of sodium carbonate.

....

Q36: Pyridinium bromide is a salt of the organic base, pyridine. A solution of pyridinium bromide has a pH of between 5 and 6. explain whether pyridine is a strong base or a weak base.

....

In order to explain the pH of a salt solution, we need to consider the equilibria involved, in particular the effect on the water equilibrium. If this equilibrium is not affected, the pH will remain the same as in pure water (pH 7). Any change in the proportions of $H^+(aq)$ and $OH^-(aq)$ ions will cause a change in pH. Three types of salt will be considered in turn.

1. salts of strong acids and strong bases;
2. salts of strong bases and weak acids;
3. salts of strong acids and weak bases.

Strong acid and strong base

Potassium nitrate solution has a pH of 7.

Potassium nitrate is the salt of a strong acid and a strong base (nitric acid and potassium hydroxide respectively). When the salt dissolves in water, the solution contains a high concentration of $K^+(aq)$ ions and $NO_3^-(aq)$ ions (from the salt) and a very low concentration of $H^+(aq)$ ions and $OH^-(aq)$ ions (from the dissociation of water).

$$H_2O\ (l) \rightleftharpoons H^+(aq) + OH^-(aq)$$

Since potassium hydroxide is strong, $K^+(aq)$ ions will not react with $OH^-(aq)$ ions. Since nitric acid is strong, $NO_3^-(aq)$ ions will not react with $H^+(aq)$ ions. The water equilibrium will not be affected. The $[H^+(aq)]$ and $[OH^-(aq]$ will remain the same as in pure water (both equal to 10^{-7} mol l^{-1}) and the solution will be neutral.

Strong base and weak acid

Sodium ethanoate solution has a pH of 9

Sodium ethanoate is the salt of a strong base and a weak acid (sodium hydroxide and ethanoic acid respectively). When the salt dissolves in water, the solution contains ethanoate ions ($CH_3COO^-(aq)$) and sodium ions ($Na^+(aq)$) from the salt and $H^+(aq)$ ions and $OH^-(aq)$ ions (from the dissociation of water). Sodium hydroxide is a strong base (100% ionised). So, if a sodium ion meets a hydroxide ion in the solution, they will not combine. However, ethanoic acid is a weak acid.

$$CH_3COOH(aq) \rightleftharpoons CH_3COO^-(aq) + H^+(aq)$$

In ethanoic acid solution, this equilibrium lies well to the left. There are lots of molecules and few ions. Remember that the same equilibrium position is reached from either direction, i.e it does not matter whether you start with 100% molecules or 100% ions. As soon as sodium ethanoate dissolves in water, the concentration of ethanoate ions is much higher than the equilibrium concentration. So ethanoate ions combine with $H^+(aq)$ ions (from water) to form ethanoic acid molecules until equilibrium is established. The removal of $H^+(aq)$ ions disrupts the water equilibrium.

$$H_2O\ (l) \rightleftharpoons H^+(aq) + OH^-(aq)$$

More water molecules dissociate in an attempt to replace the removed $H^+(aq)$ ions until a new equilibrium is established. The result is an excess of $OH^-(aq)$ ions and an alkaline pH.

Strong acid and weak base

Ammonium Chloride has a pH of 4.

Ammonium Chloride is the salt of a weak base and a strong acid (ammonia and hydrochloric acid respectively). When the salt dissolves in water the solution contains ammonium ions ($NH_4^+(aq)$ and chloride ions ($Cl^-(aq)$ from the salt and $H^+(aq)$ ions and $OH^-(aq)$ ions from the dissociation of water. Hydrochloric acid is a strong acid (100% ionised) so if a chloride ion meets a hydrogen ion in the solution they will not combine. However ammonia is a weak base:

$$NH_3(aq) \rightleftharpoons NH_4^+(aq) + OH^-(aq)$$

In ammonia solution, this equilibrium lies well to the left. There are lots of molecules and few ions. Remember that the same equilibrium position is reached from either direction, i.e it does not matter whether you start with 100% molecules or 100% ions. As soon as ammonium chloride dissolves in water, the concentration of ammonium ions is much higher than the equilibrium concentration. So ammonium ions combine with $OH^-(aq)$ ions (from water) to form ammonia molecules until equilibrium is established. The removal of $OH^-(aq)$ ions disrupts the water equilibrium.

$$H_2O\ (l) \rightleftharpoons H^+(aq) + OH^-(aq)$$

More water molecules dissociate in an attempt to replace the removed $OH^-(aq)$ ions until a new equilibrium is established. The result is an excess of $H^+(aq)$ ions and an acidic pH.

5.5 Indicators and buffers

Indicators

Acid/base indicators (or simply indicators) are weak acids which change colour depending on the pH of the solution.

HIn can be used as a general formula for an indicator and its dissociation can be represented by this equation:

$$HIn(aq) + H_2O(l) \rightleftharpoons H_3O^+(aq) + In^-(aq)$$

For a good indicator, the undissociated acid, HIn, will have a distinctly different colour from its conjugate base, In-.

The acid dissociation constant for an indicator HIn is given the symbol KIn and is represented by:

Taking the negative log of both sides gives:

$-logK_{In} = -log[H_3O^+] - log[In^-]/[HIn]$

Since $pK_{In} = -logK_{In}$

And $pH = -log[H_3O^+]$

Then $pK_{In} = pH - log[In^-]/[HIn]$

$pH = pK_{In} + log[In^-]/[HIn]$

The pH of the solution is determined by the pK_{In} of the indicator and the ratio of **[In⁻]** to **[HIn]**. Since these are different colours, the ratio of **$[In^-]$** to **[HIn]** determines the overall colour of the solution. For a given indicator, the overall colour of the solution is dependent on the pH of the solution.

[HIn] and [In-] need to differ by a factor of 10 to distinguish the colour change. The pH range for the colour change is estimated by $pK_{In} \pm 1$.

5.5.1 pH titrations

When an acid is gradually neutralised by a base, the change in pH can be monitored using a pH meter. The results can be used to produce a pH titration curve from which the **equivalence point** can be identified.

pH titration

Go online

Key point

During a titration involving a strong acid and base, there is a very rapid change in pH around the equivalence point.

In the figure below, $50cm^3$ hydrochloric acid of concentration 0.1 mol ℓ^{-1} is being neutralised by sodium hydroxide solution.

Use the graph shown as Figure 5.5, to answer the questions.

Figure 5.5: Titration curve

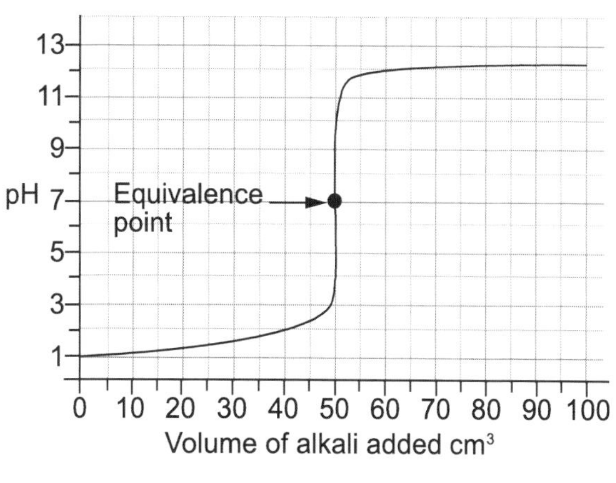

....

Q37: What is the pH at the equivalence point?

....

Q38: What is meant by the equivalence point?

....

Q39: Use the information in the graph to calculate the concentration (in mol ℓ^{-1}) of the sodium hydroxide solution.

....

Q40: Between 49.9 cm^3 and 50.0 cm^3, only 0.1 cm^3 of alkali was added. What was the change in pH for this addition?

....

Q41: Which of these statements is true?

a) The alkali is more concentrated than the acid.
b) The pH rises rapidly at the beginning.
c) The alkali is less concentrated than the acid.
d) The pH changes rapidly only around the equivalence point.

....

....

Titration curves

Figure 5.6 shows pH titration curves produced by the different combinations of strong and weak acids and alkalis. Look closely at the equivalence points in each graph.

Figure 5.6: Titration curves

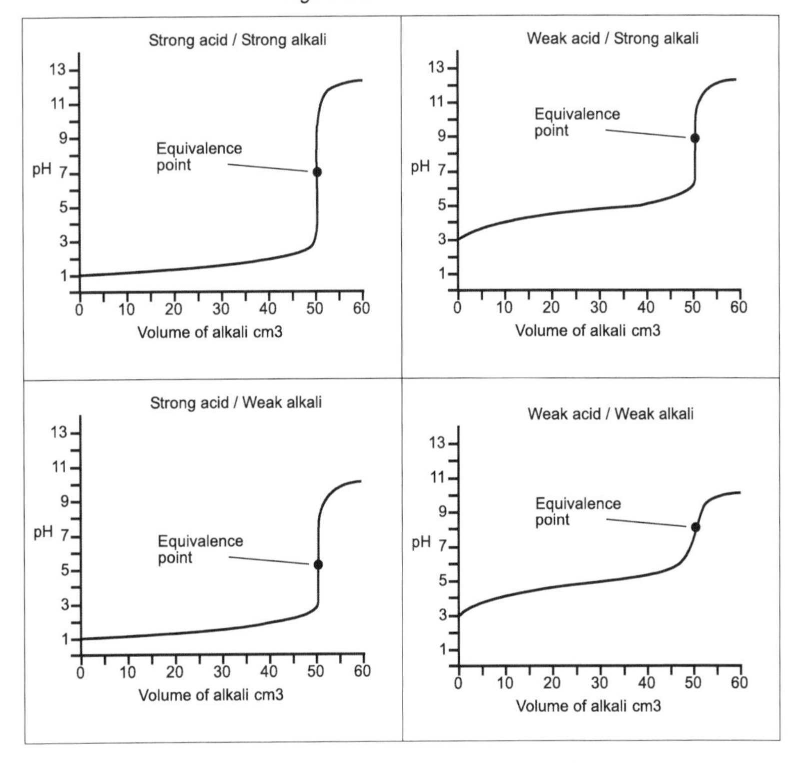

....

Q42: Which combination has an equivalence point at pH 7?

a) Strong acid/strong alkali
b) Strong acid/weak alkali
c) Weak acid/strong alkali
d) Weak acid/weak alkali

....

Q43: What is the pH of the salt formed from a strong acid and a strong alkali?

a) 1
b) 5
c) 7
d) 9

....

Q44: Which of these is the most likely pH of the salt formed from a strong acid and a weak alkali?

a) 1
b) 5
c) 7
d) 9

....

Q45: Which of these is the most likely pH of the salt formed from a weak acid and a strong alkali?

a) 1
b) 5
c) 7
d) 9

....

Q46: Using the graphs in Figure 5.6 and your previous answers, write a general statement about the pH at the equivalence point in an acid/ alkali titration. Then display the answer.

....

Since the equivalence points occur at different pH for different combinations, different indicators will be required in each case.

....

Choosing indicators

Go online

There are four different possible combinations of acid and alkali, shown in the table below.

A	strong acid/ strong alkali
B	strong acid/ weak alkali
C	weak acid/ strong alkali
D	weak acid/ weak alkali

Look at the titration curves for these four different combinations of acids and bases, then answer the questions.

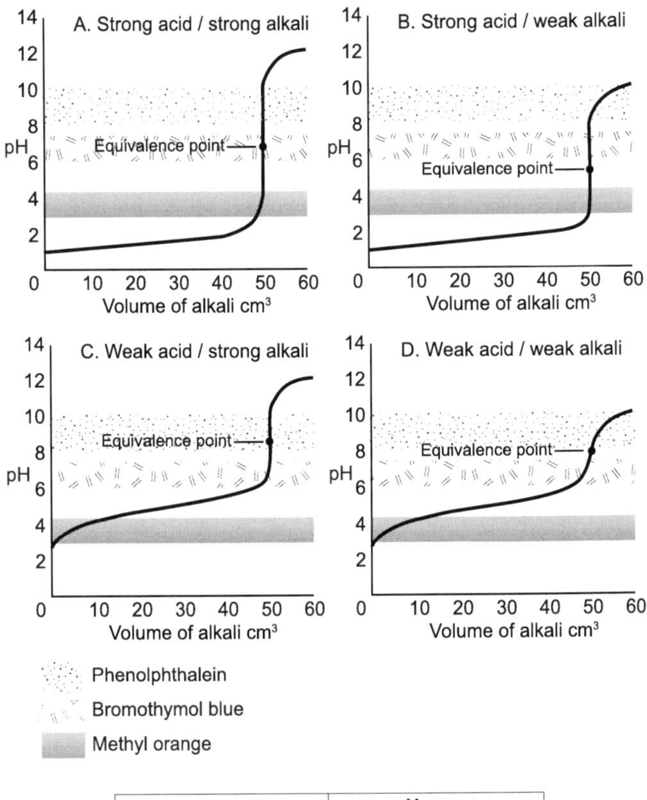

Indicator	pH range
phenolphthalein	8.2-10.0
bromothymol blue	6.0-7.6
methyl orange	3.0-4.4

Other indicators and their pH ranges are given in page 20 of the Chemistry Data Booklet (http://bit.ly/29TG6Wr).

Set A: strong acid and strong base

For each indicator, decide whether or not it is suitable for this titration

Q47: Phenolphthalein

a) Suitable
b) Unsuitable

....

Q48: Bromothymol blue

a) Suitable
b) Unsuitable

....

Q49: Methyl orange

a) Suitable
b) Unsuitable

....

Q50: Which would be the best indicator for this titration?

a) Phenolphthalein
b) Bromothymol blue
c) Methyl orange
d) None of these

....

Q51: Explain your choice.

....

Set B: Strong acid and weak base

For each indicator, decide whether or not it is suitable for this titration

Q52: Phenolphthalein

a) Suitable
b) Unsuitable

....

Q53: Bromothymol blue

a) Suitable
b) Unsuitable

....

Q54: methyl orange

a) Suitable
b) Unsuitable

....

Q55: Which would be the best indicator for this titration?

a) Phenolphthalein
b) Bromothymol blue
c) Methyl orange
d) Either methyl orange or bromothymol blue

....

Q56: Explain your choice.

....

Set C: Weak acid and strong base

For each indicator, decide whether or not it is suitable for this titration

Q57: Phenolphthalein

a) Suitable
b) Unsuitable

....

Q58: Bromothymol blue

a) Suitable
b) Unsuitable

....

Q59: Methyl orange

a) Suitable
b) Unsuitable

....

Q60: Which would be the best indicator for this titration?

a) Phenolphthalein
b) Bromothymol blue
c) Methyl orange
d) None of these

....

Q61: Explain your choice.

....

Set D: Weak acid and weak base

For each indicator, decide whether or not it is suitable for this titration

Q62: Phenolphthalein

a) Suitable
b) Unsuitable

....

Q63: Bromothymol blue

a) Suitable
b) Unsuitable

....

Q64: Methyl orange

a) Suitable
b) Unsuitable

....

Q65: Which would be the best indicator for this titration?

a) Phenolphthalein
b) Bromothymol blue
c) Methyl orange
d) None of these

....

Q66: Explain your choice.

....

....

5.5.2 Buffer solutions

Small changes in pH can have a surprisingly large effect on a system. For example, adding a small volume of lemon juice or vinegar to milk changes the protein structure and curdling occurs. Many processes, particularly in living systems, have to take place within a precise pH range. Should the pH of blood move 0.5 units outside the range shown in the table below (Table 5.1), the person would become unconscious and die. Evolution has devised buffer solutions to prevent such changes in pH in the body. A **buffer solution** is one in which the pH remains approximately constant when small amounts of acid or base are added.

Table 5.1: pH of body fluids

Fluid	pH range
blood	7.35-7.45
saliva	6.4-6.8
tears	7.4
urine	4.8-7.5
stomach juices	1.6-1.8

....

Biological systems work within precise ranges which buffers keep fairly constant. Why do you think urine can have such a wide range?

Manufacturing systems also require precise control of pH and buffers are used in electroplating, photographic work and dye manufacture. Some examples are shown below.

Many pharmacy products try to match their pH to the pH of the body tissue.

Electroplating industries need pH control over their plating solutions.

Buffer solutions are of two types:

- an acid buffer consists of a solution of a weak acid and one of its salts
- a basic buffer consists of a solution of a weak base and one of its salts

If a buffer is to stabilise pH, it must be able to absorb extra acid or alkali if these are encountered.

5.5.2.1 Acid Buffers

An acid buffer consists of a weak acid represented as **HA** . It will be slightly dissociated. Large reserves of **HA** molecules are present in the buffer.

Figure 5.7: Dissociation of a weak acid

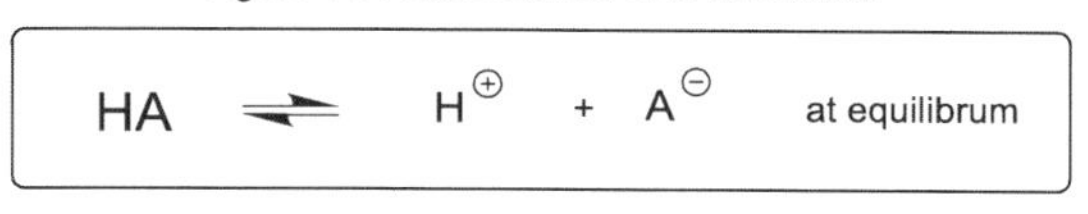

....

The weak acid salt **MA** also present will be completely dissociated. Large reserves of the **A**⁻ ion are present in the buffer. This is the conjugate base.

Figure 5.8: Dissociation of a salt of a weak acid

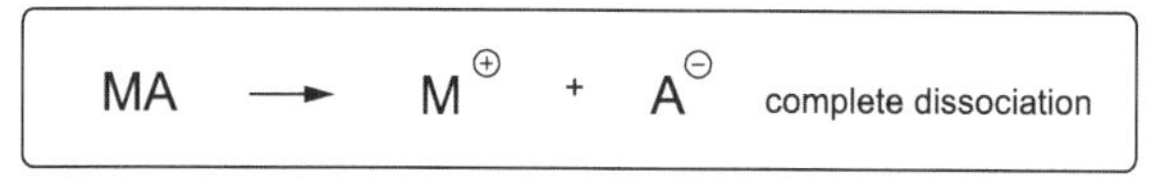

....

How does the buffer work? If acid is added to the mixture the large reserve of A^- ions will trap the extra hydrogen ions and convert them to the weak acid. This stabilises the pH. If alkali is added the large reserve of **HA** molecules will convert the extra OH^- to water. This again stabilises the pH.

A typical example of an acid buffer solution would be ethanoic acid and sodium ethanoate. The ethanoic acid is only partly dissociated. The sodium ethanoate salt completely dissociates and provides the conjugate base.

Figure 5.9: Equilibria in acid buffer solutions

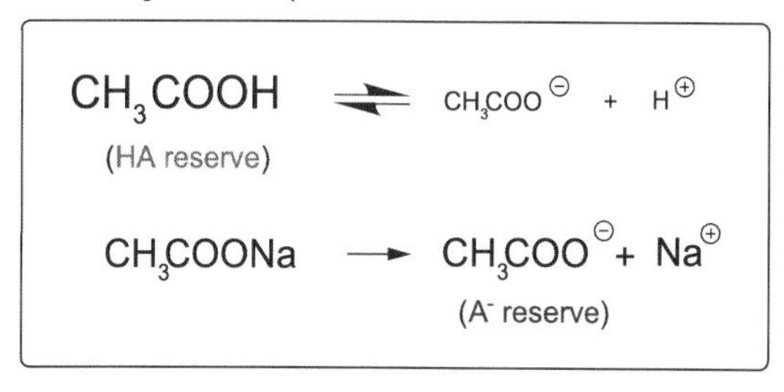

....

The stable pH of the buffer is due to:

- The weak acid which provides H^+ to trap added OH^-.
- The salt of this acid which provides A^- to trap added H^+.

Action of a buffer solution

In an acid buffer, the weak acid supplies hydrogen ions when these are removed by the addition of a small amount of base. The salt of the weak acid provides the conjugate base, which can absorb hydrogen ions from addition of small amounts af acid.

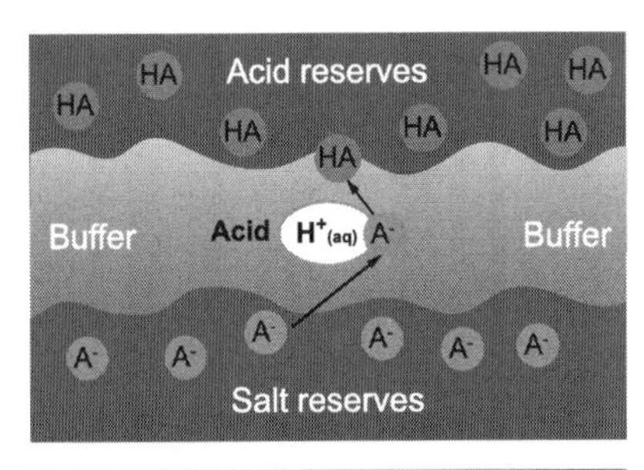

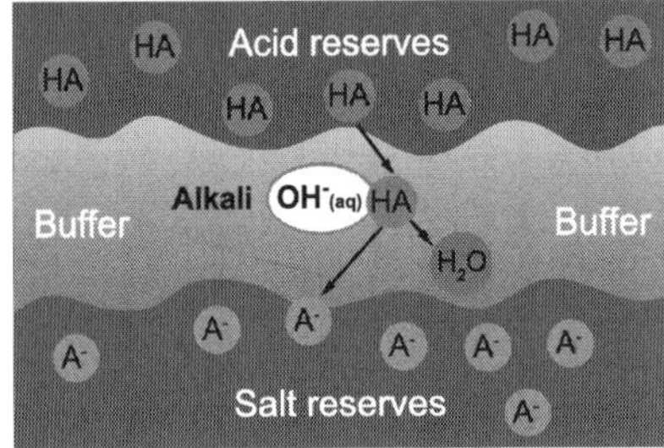

Addition of acid to the buffer

Extra hydrogen ions in the buffer upset the equilibrium situation in the weak acid.

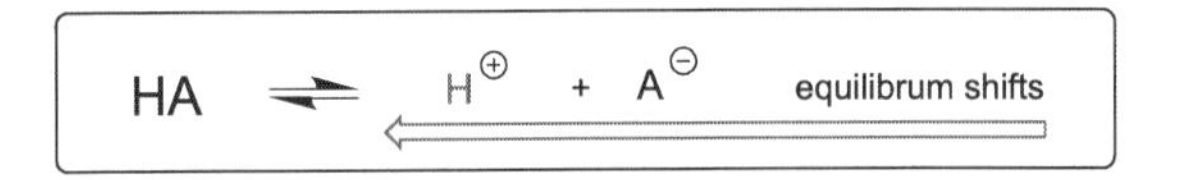

The position of equilibrium shifts (Le Chatelier's principle) and the large reserves of $\mathbf{A^-}$ ions from the salt allow the $\mathbf{H^+}$ ions to be removed. The $\mathbf{A^-}$ ions provide the conjugate base.

Addition of hydroxide to the buffer

Extra hydroxide ions in the buffer react with some $\mathbf{H^+}$ ions and upset the equilibrium situation in the weak acid.

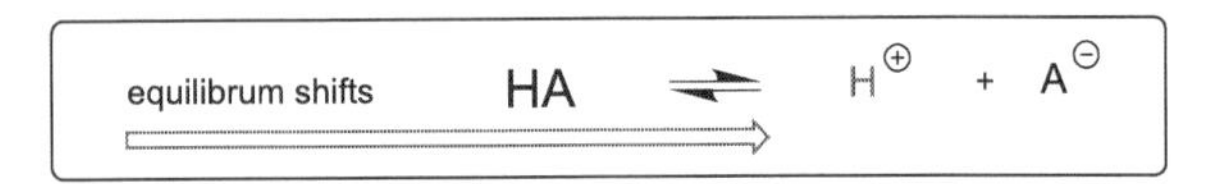

The position of equilibrium shifts (Le Chatelier's principle) and the large reserves of **HA** molecules from the weak acid allow the H^+ ions to be restored.

....

5.5.2.2 Basic buffers

A basic buffer consists of a solution of a weak base and one of its salts, e.g. ammonia solution and ammonium chloride. The ammonia solution is partly ionised and the ammonium chloride is completely ionised. If hydrogen ions are added, they combine with ammonia and if hydroxide ions are added, they combine with the ammonium ions (conjugate acid) provided by the salt (NH_4Cl).

Figure 5.10: Equilibria in base buffer solutions

$$NH_3 + H_2O \rightleftharpoons NH_4^{\oplus} + OH^{\ominus}$$

$$NH_4Cl \rightarrow NH_4^{\oplus} + Cl^{\ominus}$$

....

The stable pH of the buffer is due to:

- The weak base which provides NH_3 to trap added H^+.
- The salt of this base which provides NH_4^+ to trap added OH^-.

Summary of buffer systems

Go online

Q67: Using the word bank complete the formulae to show how a basic buffer solution made from aqueous ammonia and ammonium chloride behave when acid or alkali are added.

Adding acid H^+ + _______ $\rightleftharpoons$ _______

Adding alkali OH^- + _______ $\rightleftharpoons$ _______ + _______

Word bank

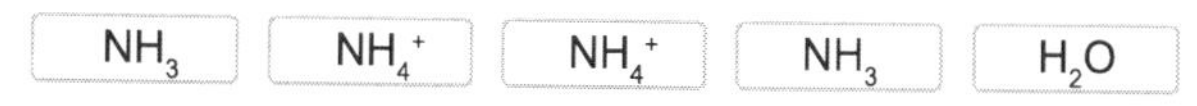

....

Q68: Using the word bank complete the formulae to show how an acidic buffer solution made from ethanoic acid and sodium ethanoate behaves when acid or alkali are added.

Adding acid H^+ + [-------] $\rightleftharpoons$ [-------]

Adding alkali OH^- + [-------] $\rightleftharpoons$ [-------] + [-------]

Word bank

CH_3COOH	CH_3COO^-	CH_3COO^-	CH_3COOH	H_2O

....

5.5.3 Calculating pH and buffer composition

A glance at Table 5.1 shows that biological buffer solutions have to operate around specific pH values. This pH value depends upon two factors. The acid dissociation constant and the relative proportions of salt and acid. The dissociation constant for a weak acid **HA** is given by this expression:

$$HA_{(aq)} \rightleftharpoons H^+_{(aq)} + A^-_{(aq)}$$

$$K_a = \frac{[H^+]\,[A^-]}{[HA]}$$

$$[H^+] = K_a \times \frac{[HA]}{[A^-]}$$

Two assumptions can be made that simplify this expression even further.

1. In a weak acid like **HA**, which is only very slightly dissociated, the concentration of **HA** at equilibrium is approximately the same as the molar concentration put into the solution.
2. The salt **MA** completely dissociates. Therefore **[A$^-$]** will effectively be the concentration supplied by the salt.

The expression becomes:

$$[H^+] = K_a \times \frac{[\text{acid}]}{[\text{salt}]}$$

taking the negative log of each side:

$$pH = pK_a - \log\frac{[\text{acid}]}{[\text{salt}]}$$

Two important points can be seen from this equation:

- Since $\frac{[\text{acid}]}{[\text{salt}]}$ is a ratio, adding water to a buffer will not affect the ratio (it will dilute each equally) and therefore will not affect the **[H$^+$]** which determines the pH.

- If the **[acid]** = **[salt]** when the buffer is made up, the pH is the same as the pK_a (or $H^+ = K_a$)

This equation allows calculation of pH of an acid buffer from its composition and acid dissociation constant, or calculation of composition from the other two values. Values for K_a and pK_a are available in the data booklet.

The next two problems show examples of the two most common type of calculation.

Example : Calculating pH from composition and K_a

Calculate the pH of a buffer solution made with 0.1 mol ℓ^{-1} ethanoic acid (K_a =1.7 × 10^{-5}) and sodium ethanoate if the salt is added:

a) at 0.1 mol ℓ^{-1}

b) at 0.2 mol ℓ^{-1}

a) With the salt at 0.1 mol ℓ^{-1}

$$\left[H^+\right] = K_a \times \frac{[\text{acid}]}{[\text{salt}]}$$
$$\left[H^+\right] = 1.7 \times 10^{-5} \times \frac{0.1}{0.1}$$
$$\left[H^+\right] = 1.7 \times 10^{-5} mol\ l^{-1}$$
$$pH = -\log\left[H^+\right] = -\log\left(1.7 \times 10^{-5}\right)$$
$$pH = -(0.23 - 5.00)$$
$$pH = 4.77$$

b) With the salt at 0.2 mol ℓ^{-1}

$$\left[H^+\right] = K_a \times \frac{[\text{acid}]}{[\text{salt}]}$$
$$\left[H^+\right] = 1.7 \times 10^{-5} \times \frac{0.1}{0.2}$$
$$\left[H^+\right] = 1.7 \times 10^{-5} \times 0.5\ mol\ l^{-1}$$
$$pH = -\log\left[H^+\right] = -\log\left(8.5 \times 10^{-6}\right)$$
$$pH = -(0.93 - 6.00)$$
$$pH = 5.07$$

....

Notice that doubling the salt concentration has only raised the pH by 0.3 (from 4.76 to 5.07) and in general, the pH of the buffer is tied closely to the pK_a value for the weak acid, in this case ethanoic acid pK_a = 4.76. The ratio of acid to salt effectively provides a 'fine tuning' of the pH.

Example : Calculating composition from pH and pK_a

Calculate the concentrations of ethanoic acid and sodium ethanoate required to make a buffer solution with a pH of 5.3 (pK_a in data booklet).

So the ratio of **3.47 to 1** is required and 3.47 moles of sodium ethanoate mixed with one litre of 1.0 mol ℓ^{-1} ethanoic acid could be used.

....

Go online

Buffer calculations

Q69: Calculate the pH of a buffer solution containing 0.10 mol ℓ^{-1} ethanoic acid and 0.50 mol ℓ^{-1} sodium ethanoate (K_a is in the data booklet). Give your answer to two decimal places.

....

Q70: Calculate the composition of methanoic acid and sodium methanoate required to make a buffer solution with a pH of 4.0. Quote your answer as a ratio of salt to 1 (so 6.31 to 1 would quote as 6.31).

....

Q71: A 0.10 mol ℓ^{-1} solution of a weak acid has 0.40 mol ℓ^{-1} of its sodium salt dissolved in it. The resulting buffer has a pH 5.35. Find the dissociation constant of the acid.

....

Q72: What pH (to one decimal place) would be expected if 7.20g of sodium benzoate was dissolved in one litre of 0.02 mol ℓ^{-1} benzoic acid (sodium benzoate is C_6H_5COONa, K_a and pK_a in data booklet).

....

Q73: To prepare 1 litre of a buffer solution which would maintain a pH 5.5, 0.6g of ethanoic acid was used. What mass in grams of sodium ethanoate should the solution contain? Answer to two decimal places.

....

Q74: One of the systems which maintains the pH of blood at 7.40 involves the acid $H_2PO_4^-$ and the salt containing HPO_4^{2-}. Calculate the ratio of salt to acid in blood (K_a and pK_a of $H_2PO_4^-$ in data booklet). Assume that the weak acid is $H_2PO_4^-$ and the salt contains HPO_4^{2-}.

....

....

Extra questions

Q75: Ethanoic acid CH_3COOH is a weak acid. What is the conjugate base of ethanoic acid?

a) CH_3
b) OH^-
c) CH_3COO^-
d) H^+

....

Q76: What is the conjugate base of H_2SO_4?

a) H^+
b) OH^-
c) SO_4^{2-}
d) HSO_4^-

....

Q77: Lithium hydroxide is a strong base. What is the conjugate acid of lithium hydroxide?

a) Li^+
b) OH^-
c) H^+
d) H_2O

....

Q78: Is a H^+ concentration of 10^{-2} mol ℓ^{-1} greater or smaller than 10^{-12} mol ℓ^{-1}?

a) Greater
b) Smaller

....

Q79: As the concentration of the H^+ ion increases the value of the pH:

a) Increase.
b) Decreases.

....

Calculate the pH of the following solutions to two decimal places.

Q80: A sample of water with $[H^+] = 1.0 \times 10^{-7}$ mol ℓ^{-1}

....

Q81: A soft drink with $[H^+] = 3.1 \times 10^{-4}$ mol ℓ^{-1}.

....

Q82: A blood sample with $[H^+] = 4.0 \times 10^{-8}$ mol ℓ^{-1}

....

Exam hint:
Be careful to press the **log** or $\mathbf{log_{10}}$ key on your calculator to find the pH. Do not use the **ln** or $\mathbf{log_e}$ key.

Key point

The pH scale shows whether a solution is acidic or alkaline.
Solutions for which $pH < 7$ are acidic; pH 7 is neutral; and $pH > 7$ are alkaline.

The dissociation constant of water, K_w

Pure water dissociates as shown to a small extent:

$$H_2O(l) \rightleftharpoons H^+(aq) + OH^-(aq)$$

Q83: Equilibrium constant is given by which expression?

a) $Kc = \frac{[H^+]\,[OH^-]}{[H_2O]}$

b) $Kc = \frac{[H_2O]\,[H^+]}{[OH^-]}$

c) $Kc = \frac{[OH^-]\,[H_2O]}{[H^+]}$

. . . .

In this expression, $[H_2O]$ is effectively constant.

Q84: The ionic product of water with the special symbol K_w is given by which expression?

a) $K_w = [H_2O]$
b) $K_w = [H_2O] / [H^+] [OH^-]$
c) $K_w = [H^+] [OH^-]$
d) $K_w = [H^+] [OH^-] / [H_2O]$

. . . .

Q85: Using the word bank below complete the following sentences.

Word Bank: neutral water; greater than; less than; -2; -7; -14.

In —— the concentrations of the ionic species $[H^+]$ is equal to $[OH^-]$ and the value of K_w is 1.0×10 —— at 25°C.

In acidic solutions the concentrations of the ionic species $[H^+]$ is —— $[OH^-]$ and the value of K_w is 1.0×10 —— at 25°C.

In alkaline solution the concentrations of the ionic species $[H^+]$ is —— $[OH^-]$ and the value of K_w is 1.0×10 —— at 25°C.

. . . .

The acid dissociation constant, K_a and pK_a

Some acids in water dissociate only partially to form H^+ ion and the conjugate base. These acids are called weak acids.
For a weak acid HA the equilibrium in solution is described by:

$$HA(aq) \rightleftharpoons H^+(aq) + A^-(aq)$$

where A^- is the conjugate base.

The equilibrium constant K_a - called the acid dissociation constant - is

$$K_a = \frac{[H^+][A^-]}{[HA]}$$

As with $[H^+]$ and pH, the term pK_a has been invented to turn small numbers into more manageable ones.

$$pK_a = -\log_{10}K_a$$

Q86: The acid dissociation constant K_a of nitrous acid is 4.7×10^{-4}. What is the value of pK_a? Answer to one decimal place.

....

Q87: An acid has a pK_a of 9.8. What is the value of K_a?

....

Q88: The table lists the K_a values for four acids. Write the letters of these acids in order of decreasing acid strength (strongest first). Do not leave spaces between the four letters in your answer.

Letter	Name	K_a
A	Alanine	9.0×10^{-10}
B	Butanoic acid	1.5×10^{-5}
C	Carbonic acid	4.5×10^{-7}
D	Dichloroethanoic acid	5.0×10^{-2}

....

Q89: The table lists the pK_a values for four acids. Write the letters of these acids in order of decreasing acid strength (strongest first). Do not leave spaces between the four letters.

Letter	Name	pK_a
A	Benzoic acid	4.2
B	Boric acid	9.1
C	Bromoethanoic acid	2.9
D	Butanoic acid	4.8

....

Buffer solutions

A buffer solution is one that resists a change in pH when moderate amounts of an acid or base are added to it.

Let us imagine we add small amounts of acid to water. The pH of the water to begin with is 7 as it is neutral. The table shows the pH of a series of solutions prepared from adding successive 2 cm^3 portions of 0.1 mol ℓ^{-1} acid to 100 cm^3 of water.

Volume of acid added/cm^3	0	2	4	6	8	10
pH	**7.00**	2.71	2.42	2.25	2.13	2.04

The pH drops a lot and the solution gets more acidic. Note particularly the drop in pH from 7.00 to 2.71 on adding the first 2 cm^3 of acid.

Now consider the next table where the same volumes of acid are added to 100 cm^3 of a buffer solution. The buffer solution was made up to have an initial pH of 5.07 using a mixture of ethanoic acid and sodium ethanoate in water.

Volume of acid added/cm^3	0	2	4	6	8	10
pH	**5.07**	5.06	5.05	5.03	5.02	5.01

The data show that the pH of the buffer solution changes only slightly with the addition of the acid. Compared with water, it drops by only 0.01 pH units after the first 2 cm^3 of acid is added.

Buffers are very useful because of this property. For many industrial processes and for biological situations it is important to keep a constant pH. For example, buffers exist in human blood, otherwise a change in pH of 0.5 unit might cause you to become unconscious.

The same number of H^+ ions were added to both the water and the buffer solution. What happened to most of the H^+ ions added to the buffer? How does a buffer work? The secret lies in the mixture of ingredients. The acid buffer described above is a solution of sodium ethanoate ($CH_3COO^- Na^+$) and ethanoic acid (CH_3COOH) in water.

Q90: With this combination is the pH acidic or basic? Type "acidic" or "basic".

....

Q91: Will the pH be less than or greater than 7? Answer "less" or "greater".

....

Ethanoic acid is a weak acid and sodium ethanoate is the sodium salt of ethanoic acid. Sometimes this is called the conjugate base. But more about that later.

A basic buffer is a solution of a weak base such as ammonia and the corresponding salt such as ammonium chloride.

Q92: Will the pH be less than or greater than 7? Answer "less" or "greater".

....

Key point

Depending on the composition of the buffer solution, the solution can be acidic (pH<7), neutral (pH=7) or basic (pH>7).

Returning to an acidic buffer such as sodium ethanoate/ethanoic acid solution, remember that sodium ethanoate is a salt and dissociates completely in aqueous solution;

$$CH_3COO^-Na^+(aq) \rightarrow CH_3COO^-(aq) + Na^+(aq)$$

to produce ethanoate ions.

Ethanoic acid, because it is a weak acid, dissociates only slightly;

$$CH_3COOH(aq) \rightleftharpoons CH_3COO^-(aq) + H^+(aq)$$

and reaches an equilibrium where most of the ethanoic acid exists as undissociated molecules. Because, in the buffer solution, ethanoate ions are provided by sodium ethanoate, the equilibrium position shifts to form even more ethanoic acid molecules.

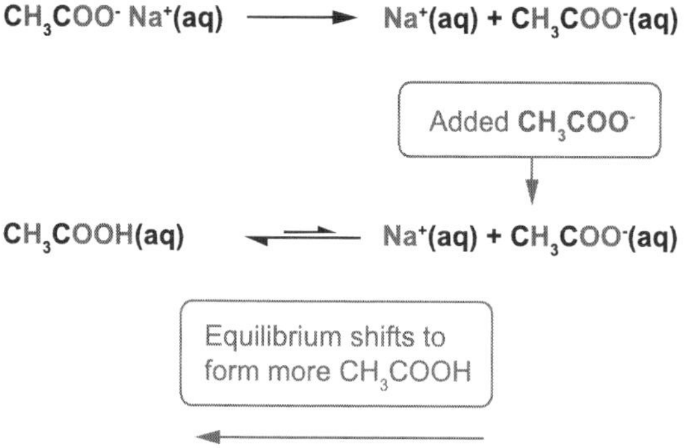

$$\mathbf{CH_3COO^-\,Na^+(aq) \longrightarrow Na^+(aq) + CH_3COO^-(aq)}$$

Added $\mathbf{CH_3COO^-}$

$$\mathbf{CH_3COOH(aq) \rightleftharpoons Na^+(aq) + CH_3COO^-(aq)}$$

Equilibrium shifts to form more CH_3COOH

The result is that the buffer solution contains a reservoir of ethanoic acid molecules and ethanoate ions. As we shall see, this gives it the capability of reacting with either acid or base added to the buffer solution thus minimising their effect.

Addition of acid (H^+)

When H^+ is added to the buffer solution, the H^+ reacts with the CH_3COO^- ions (present mainly from the sodium ethanoate) to form ethanoic acid. On account of this reaction, the H^+ ion concentration remains virtually unchanged; the pH hardly alters.

$$CH_3COO^-(aq) + H^+(aq) \rightarrow CH_3COOH(aq)$$

Addition of alkali (OH^-)

When OH^- ion is added to the buffer solution, the OH^- reacts with CH_3COOH (base plus acid forms a salt plus water);

$$CH_3COOH(aq) + OH^-(aq) \rightarrow CH_3COO^-(aq) + H_2O(l)$$

to form CH_3COO^- and, because the extra OH^- is removed, the pH of the buffer solution hardly changes.

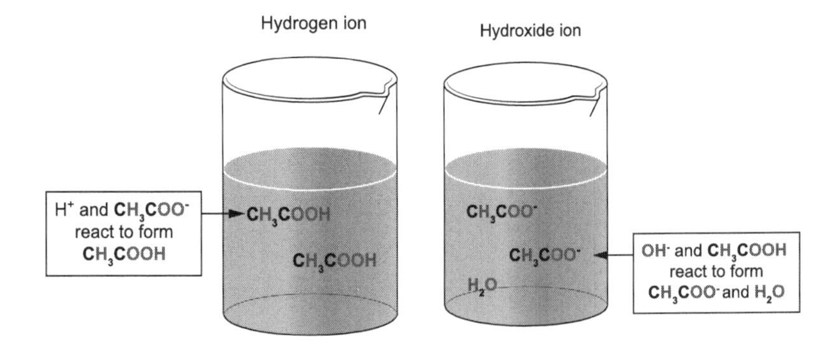

Conjugate acids and bases

There are a large number of buffer solutions. One was given as an example in this section. In general, one component of an acid buffer is a weak acid represented as **HA**. It will be only slightly dissociated. Large reserves of **HA molecules** are present in the buffer.

$$HA \rightleftharpoons H^{\oplus} + A^{\ominus} \quad \text{at equilibrum}$$

The other component is a salt **MA** of the weak acid. This will be completely dissociated. Large reserves of the **A^- ion** are present in the buffer. This is the **conjugate base**.

$$MA \rightarrow M^{\oplus} + A^{\ominus}$$ complete dissociation

Consider an acidic buffer consisting of a solution of phosphoric acid (H_3PO_4) and potassium dihydrogenphosphate ($K^+H_2PO_4^-$)

Q93: When H^+ is added to this buffer, which component will react with it?

a) H_3PO_4
b) $H_2PO_4^-$

....

Q94: What will form from this reaction?

....

Now let us consider a buffer containing a weak base and a salt.

The figure shows the two reactions relevant to a buffer solution containing ammonia and ammonium chloride, where ammonia is the weak base.

$NH_4^+Cl^-(aq) \rightarrow NH_4^+(aq) + Cl^-(aq)$

Added NH_4^+

$NH_3(aq) + H^+(aq) \rightleftharpoons NH_4^+(aq)$

Equilibrium shifts to form more NH_3

Q95: Is this buffer acidic or basic?

a) Acidic
b) Basic

....

Q96: Name the base in a buffer solution containing ammonia and ammonium chloride.

....

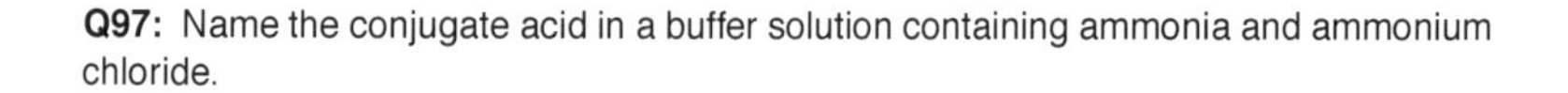

Q97: Name the conjugate acid in a buffer solution containing ammonia and ammonium chloride.

....

Q98: When acid is added to this buffer solution the H^+ combine with:

a) Ammonia, NH_3,
b) Ammonium ion, NH_4^+ ,
c) Ammonium chloride, NH_4Cl?

....

Q99: When alkali is added to this buffer solution the OH^- combine with:

a) Ammonia, NH_3,
b) Ammonium ion, NH_4^+,
c) Ammonium chloride, NH_4Cl?

....

Q100: Using the word bank complete the paragraph.

Word bank: salts; acidic; weak base; weak acid; ammonia; ethanoic acid; basic.

An —— buffer solution contains a mixture of a —— and one of its ——. An example is a mixture of —— and potassium ethanoate in water.

A —— buffer solution contains a mixture of a —— and one of its ——. An example is a mixture of —— and ammonium chloride in water.

....

Key point

An acid buffer contains a weak acid and a salt of the same weak acid.
A basic buffer contains a weak base and a salt of the same weak base.

Calculating the pH of buffer solutions

In order to calculate the pH of an acidic buffer solution we need to know:

- the acid dissociation constant K_a
- the relative concentrations of the acid and its salt

For an equilibrium reaction of ethanoic acid:

$$CH_3COOH(aq) \rightleftharpoons CH_3COO^-(aq) + H^+(aq)$$

the equilibrium constant expression K_a is:

$$K_a = \frac{[H^+][CH_3COO^-]}{[CH_3COOH]}$$

Note that the square brackets represent concentrations in moles per litre.

Calculate the molecular mass of ethanoic acid.

Q101:

Relative formula mass of ethanoic acid.

Enter the correct values from the SQA booklet into the boxes below.

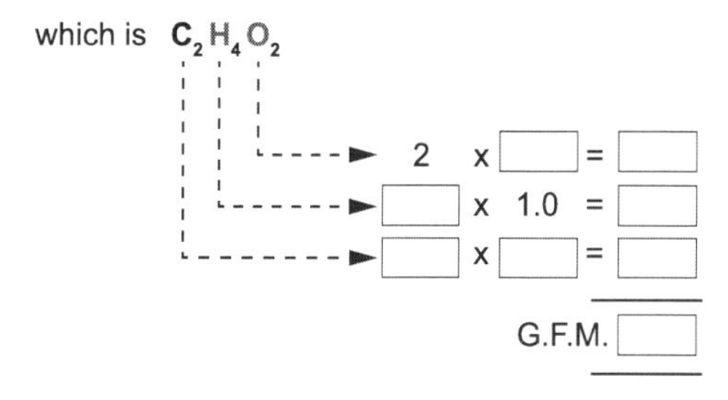

. . . .

This number of grams when made up to 1 litre of aqueous solution will have a concentration of 1 mole per litre.

Q102: How many grams of ethanoic acid would be required to dissolve in 1 litre of solution to make a 0.1 mol ℓ^{-1} solution? i.e. $[CH_3COOH]$ = 0.1 Answer to 1 decimal place without units.

. . . .

Q103: At equilibrium at 25°C, the concentrations of each species in the 0.1 mol ℓ^{-1} solution are:

$[CH_3COO^-] = 1.304 \times 10^{-3}$, $[H^+] = 1.304 \times 10^{-3}$

and $[CH_3COOH] = 0.100 - 0.0013 = 0.0987$

Note that whilst we dissolved 0.1 moles of CH_3COOH in water only a small fraction of the molecules have dissociated to form CH_3COO^- and H^+.

Calculate K_a for ethanoic acid at 25°C . Answer in standard form to 3 significant figures.

. . . .

The formula below shows you the derivation of an equation to calculate the pH of a buffer solution. You will use this equation in the remaining questions in this section.

The formula below shows you the derivation of an equation to calculate the pH of a buffer solution. You will use this equation in the remaining questions in this section.

For the dissociation of an acid HA

$$HA \rightleftharpoons H^+ + A^-$$

The equilibrium constant K_a is

$$K_a = \frac{[H^+][A^-]}{[HA]}$$

Rearranging to make $[H^+]$ the subject

$$[H^+] = K_a \frac{[HA]}{[A^-]}$$

Taking negative log_{10}

$$-log_{10}[H^+] = -log_{10}K_a - log_{10}\frac{[HA]}{[A^-]}$$

Substituting $-log_{10}[H^+] = pH$ and $-log_{10}K_a = pK_a$

$$pH = pK_a - log_{10}\frac{[HA]}{[A^-]}$$

Mathematical relationships

Table 5.2: pH of buffers

[acid]/[salt]	10:1	2:1	**1:1**	1:2 (0.5:1)	1:10 (0.1:1)
log_{10}[acid]/[salt]	1.0	0.30	**0**	-0.30	-1.0
pH	pK_a - 1.0	pK_a - 0.30	**pK_a**	pK_a + 0.30	pK_a + 1.0

....

The table shows that fine tuning of the pH of a buffer solution can be achieved by altering the ratio of [acid]/[salt]. For example, the pK_a of propanoic acid is 4.87 and thus the pH of a propanoic acid/sodium propanoate buffer would be 4.87 for a ratio of [acid]/[salt] = 1:1. (pH = pK_a)

Q104: A buffer solution is prepared using 0.1 mol ℓ^{-1} ethanoic acid and 0.2 mol ℓ^{-1} sodium ethanoate. Calculate the pH to two decimal places.

....

Q105: A buffer solution is prepared using 0.2 mol ℓ^{-1} ethanoic acid and 0.1 mol ℓ^{-1} sodium ethanoate. Calculate the pH to two decimal places.

....

Q106: A buffer consists of a aqueous solution of boric acid (H_3BO_3) and sodium borate (NaH_2BO_3). Use your data booklet to find the pK_a of boric acid. If the concentration of the acid was 0.2 mol ℓ^{-1} and the pH of the buffer solution was 9.4, what was the concentration of the sodium borate?

....

Try these two questions to ensure that you understand buffers.

Q107: You add 1.0 cm^3 of 0.2 mol ℓ^{-1} HCl to each of the following solutions. Which one will show the least change of pH?

a) 100 cm^3 of 0.1 mol ℓ^{-1} HCOOH
b) 100 cm^3 of 0.1 mol ℓ^{-1} HCOOH/0.2 mol ℓ^{-1} $HCOO^-Na^+$
c) 100 cm^3 of water

....

Q108: A buffer solution is prepared using 0.1 mol ℓ^{-1} ethanoic acid and 0.2 mol ℓ^{-1} sodium ethanoate. 1 cm^3 of the buffer solution is diluted with water to a final volume of 10 cm^3. Afterwards, does the pH increase, decrease or stay the same?

a) Increase
b) Decrease
c) Stay the same

....

....

5.6 Summary

Summary

You should now be able to:

- describe the equilibrium chemistry of acids and bases, write equilibrium expressions;
- describe the chemistry of acids and bases;
- use the terms: pH, Kw, Ka and pKa;
- understand the chemistry of buffer solutions;
- calculate the pH of buffer solutions.

5.7 Resources

- Acids and bases (http://www.nclark.net/AcidsBases)
- pH Rainbow Tube (https://youtu.be/ZT9Ie3AUI7E)
- http://www.chemguide.co.uk (ChemConnections equilibrium)
- Davidson virtual chemistry experiments (http://bit.ly/29PGWyy)
- Chemistry tutorial 11.4a Bronsted Lowry Theory Of Acids And Bases (https://youtu.be/fM52LrQmel0)
- Bronsted Lowry Theory of Acids and Bases (https://youtu.be/0O2GekIXIoA)
- Iowa State University Department of Chemistry (http://chem.iastate.edu)

5.8 End of topic test

End of Topic 5 test

Go online

The end of topic test for *Chemical equilibrium.*

Q109: The value of K_c for a reaction at 500 K is 0.02. At 1000 K, it is 0.10.

Which of the following statements is true?

a) The reaction is endothermic.
b) The reaction is exothermic.
c) $\Delta H = 0$
d) The reaction is 5 times as fast at 1000 K as at 500 K.

. . . .

Q110: Which of the following could have an equilibrium constant equal to 1×10^{-55}?

a) $HCl (aq) + NaOH (aq) \rightleftharpoons NaCl (aq) + H_2O (l)$
b) $Cu (s) + Mg^{2+} (aq) \rightleftharpoons Cu^{2+} (aq) + Mg (s)$
c) $Zn (s) + 2Ag^{+} (aq) \rightleftharpoons Zn^{2+} (aq) + 2Ag (s)$
d) $CH_3OH (l) + CH_3COOH (l) \rightleftharpoons CH_3COOCH_3 (l) + H_2O (l)$

. . . .

Q111: Which of the following will increase the equilibrium constant for the following reaction given that ΔH, left to right, is positive?

$N_2O_4(g) \rightleftharpoons 2NO_2(g)$

a) Increase of pressure
b) Decrease of temperature
c) Use of a catalyst
d) Increase of temperature

. . . .

Q112: In the reaction:

$3O_2(g) \rightleftharpoons 2O_3(g)$

K_p is 1×10^{-4}

What is the partial pressure of ozone, (O_3), if that of oxygen is 4 atm?

a) 2.67 atm
b) 8×10^{-2} atm
c) 4×10^{-4} atm
d) 1×10^{-4} atm

. . . .

The Haber process is represented by the equation:

N_2 (g) + $3H_2$ (g) $\rightleftharpoons$ $2NH_3$ (g)

for which ΔH° = -92 kJ mol^{-1}

2.0 moles of each reactant were allowed to react and come to equilibrium in a 1 litre container at 400 K. At equilibrium, 0.4 moles of ammonia were formed.

Q113: Which equilibrium constant expression is correct for this reaction?

a) $\dfrac{[H_2][N_2]}{[NH_3]}$

b) $\dfrac{[NH_3]}{[H_2][N_2]}$

c) $\dfrac{[NH_3]^2}{[H_2]^3[N_2]}$

d) $\dfrac{[H_2]^3[N_2]}{[NH_3]^2}$

....

Q114: Calculate the equilibrium concentration of nitrogen.

....

Q115: Calculate the equilibrium concentration of hydrogen.

....

Q116: Calculate the value of the equilibrium constant at this temperature.
K =

....

Q117: Explain what will happen to the value of K, if the temperature is now raised to 600 K.

....

Q118: When a solute is shaken into two immiscible liquids, it partitions itself between the two liquids in a definite ratio.

The value of this ratio is NOT dependent on the:

a) solute type.
b) temperature.
c) immiscible liquids involved.
d) mass of solute involved.

....

Q119: When a solute is distributed between two immiscible liquids and a partition coefficient of 2.0 is reached, the exchange rate between the two layers would be:

a) in a ratio of 2:1.
b) in a ratio of 1:2.
c) zero.
d) equal.

....

Q120: This diagram shows twelve moles of solute distributed between two immiscible liquids (one sphere represents one mole).

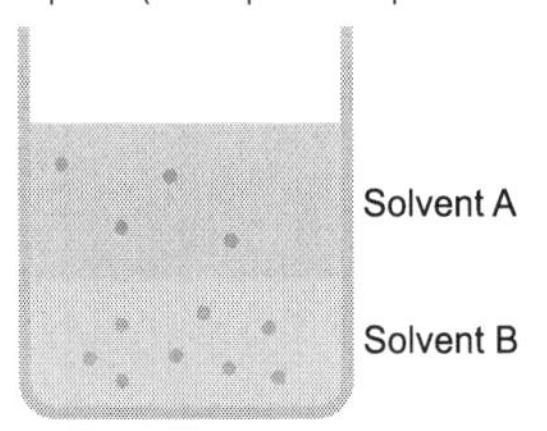

What would be the value of the partition coefficient solventA/solventB?

a) 0.5
b) 2
c) 4
d) 12

....

Q121: An aqueous solution of a monoprotic organic acid is shaken with ether at 25 °C until equilibrium is established.

20 cm^3 of the ether layer requires 15 cm^3 of 0.020 mol l^{-1} potassium hydroxide to neutralise and 20 cm^3 of the aqueous layer requires 7.5 cm^3 of 0.020 mol l^{-1} potassium hydroxide to neutralise.

Calculate the concentration of the organic acid in the ether layer.

....

Q122: Calculate the concentration of the organic acid in the aqueous layer.

....

Q123: Calculate the partition coefficient (ether/water) at 25°C.
K =

....

Q124: An aqueous extract from a normal animal feedstuff and another from a feedstuff contaminated with a toxin were subjected to thin layer chromatography. The result is shown here.

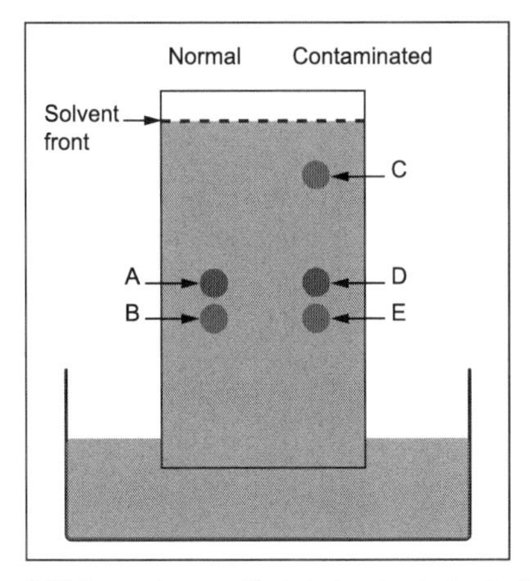

Which spot is most likely to be due to the toxin?

a) A
b) B
c) C
d) D
e) E

....

Q125: To prepare some toxin for further tests, a 50 cm^3 sample of the aqueous extract from the contaminated feedstuff was extracted with 50 cm^3 of hexane.

Using the information from the thin layer chromatography, which of the spots in the contaminated feedstuff is the least polar?

a) A
b) B
c) C
d) D
e) E

....

Q126: Into which layer would the toxin partition?

a) Water
b) Hexane

....

Q127: In which reaction is water behaving as a Bronsted-Lowry acid?

a) H_2O (l) + HF (aq) → H_3O^+ (aq) + F^- (aq)
b) H_2O (l) + NH_3 (aq) → NH_4^+ (aq) + OH^- (aq)
c) H_2O (l) + H_2SO_4 (l) → H_3O^+ (aq) + HSO_4^- (aq)
d) $2H_2O$ (l) + $2e^-$ → $2OH^-$ (aq) + H_2 (g)

....

Q128: For the reaction

NH_3 (aq) + HCO_3^- (aq) → NH_4^+ (aq) + CO_3^{2-} (aq)

which of the following statements is correct?

a) NH_3 is the conjugate acid of NH_4^+.
b) HCO_3^- is the conjugate base of CO_3^{2-}.
c) CO_3^{2-} is the conjugate acid of HCO_3^- .
d) NH_4^+ is the conjugate acid of NH_3.

....

Q129: A solution of sulfuric acid is diluted until it has a concentration of 1×10^{-5} mol l^{-1}.

Assuming that the acid is completely dissociated into ions, the pH would then be:

a) 1
b) 4.3
c) 4.7
d) 5

....

Q130: Sulfurous acid (H_2SO_3) is diprotic and dissociation occurs in two successive steps.

H_2SO_3 (aq) $\rightleftharpoons$ H^+ (aq) + HSO_3^- (aq) $K_a = 1.5 \times 10^{-2}$

HSO_3^- (aq) $\rightleftharpoons$ H^+ (aq) + SO_3^{2-} (aq) $K_a = 6.2 \times 10^{-8}$

Identify the **false** statements below referring to the hydrogensulfite ion.

a) HSO_3^- is a stronger base than SO_3^{2-}.
b) HSO_3^- is the conjugate acid of SO_3^{2-}.
c) HSO_3^- is a weaker acid than H_2SO_3.
d) pK_a for HSO_3^- is less than pK_a for H_2SO_3.
e) HSO_3^- is amphoteric.
f) HSO_3^- is the conjugate base of H_2SO_3.

....

Q131: K_w is the ionic product of water.

Which of the following expressions correctly represents K_w?

a) $K_w = - \log ([H^+] [A^-])$
b) $K_w = [H^+] [OH^-]$
c) $$K_w = \frac{[H^+][A^-]}{[HA]}$$
d) $K_w = - \log [H^+]$

....

Q132: The dissociation constant for water (K_w) varies with temperature.

$K_w = 0.64 \times 10^{-14}$ at 18°C

$K_w = 1.00 \times 10^{-14}$ at 25°C

From this information we can deduce that:

a) the ionisation of water is exothermic.
b) only at 25°C does the concentration of H^+ equal the concentration of OH^-.
c) the pH of water is greater at 25°C than at 18°C.
d) water will have a greater electrical conductivity at 25°C than at 18°C.

....

Q133: 5.0 cm^3 of a solution of hydrochloric acid was diluted to exactly 250 cm^3 with water. The pH of this diluted solution was 2.0.

The concentration of the original undiluted solution was:

a) 0.02 mol l^{-1}
b) 0.50 mol l^{-1}
c) 0.40 mol l^{-1}
d) 0.04 mol l^{-1}

....

Q134: A solution with a pH of 3.2 has a **hydroxide** ion concentration which lies between:

a) 10^{-11} and 10^{-12} mol l^{-1}
b) 10^{-3} and 10^{-4} mol l^{-1}
c) 10^{-10} and 10^{-11} mol l^{-1}
d) 10^{-2} and 10^{-3} mol l^{-1}

....

Q135: Using information from the SQA data booklet, calculate the pH of a solution of ethanoic acid of concentration 0.01 mol l^{-1}.

....

Q136: A 0.05 mol l^{-1} solution of chloroethanoic acid has a pH of 2.19. Use this information to calculate pK_a for chloroethanoic acid to 2 significant figures.

....

Q137: Indicators are weak acids for which the dissociation can be represented by:

$$HIn\ (aq) + H_2O\ (l) \rightleftharpoons H_3O^+\ (aq) + In^-\ (aq)$$

Colour 1 Colour 2

For any indicator, HIn, in aqueous solution, which of the following statements is correct?

a) The overall colour of the solution depends only on the pH.
b) In acidic solution, colour 2 will dominate.
c) The overall colour of the solution depends on the ratio of [HIn] to [In^-].
d) Adding alkali changes the value of K_{in}.

....

Q138: An indicator (H_2A) is a weak acid, and undergoes a two stage ionisation. The colours of the species are shown.

Stage 1 Stage 2

$$H_2A \rightleftharpoons HA^- + H^+ \rightleftharpoons A^{2-} + 2H^+$$

Yellow Blue Green

The dissociation constants for the two ionisations are given by:

$pK_1 = 3.5$ and $pK_2 = 5.9$.

Given that for an indicator pK = pH at the point where the colour change occurs, the indicator will be:

a) blue in a solution of pH 3 and green in a solution of pH 5.
b) yellow in a solution of pH 3 and blue in a solution of pH 5.
c) yellow in a solution of pH 3 and green in a solution of pH 5.
d) blue in a solution of pH 3 and blue in a solution of pH 5.

....

Q139: Which of the following graphs represents the change in pH as a strong alkali is added to a weak acid?

a)

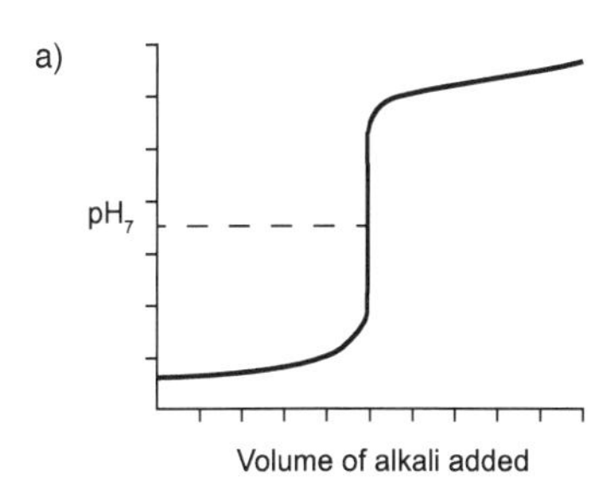

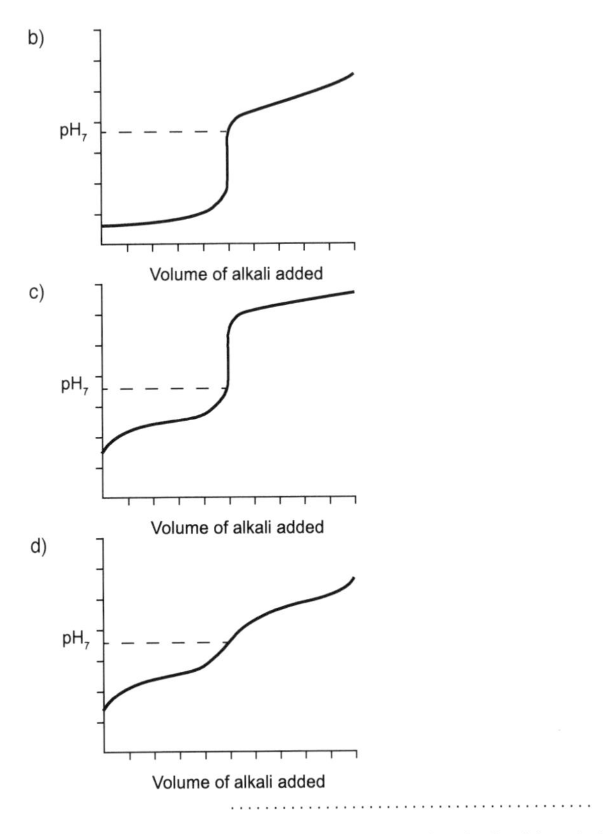

....

Q140: A titration was carried out using potassium hydroxide solution and ethanoic acid. Which is the most suitable indicator for this titration?

a) Universal indicator, pH of colour change 4.0-11.0
b) Bromothymol blue indicator, pH of colour change 6.0-7.6
c) Phenolphthalein indicator, pH of colour change 8.0-9.8
d) Methyl orange indicator, pH of colour change 3.0-4.4

....

Q141: Why is it not practical to find the concentration of a solution of ammonia by titration with standard propanoic acid solution using an indicator?

a) The pH changes gradually around the equivalence point.
b) The salt of the above acid and alkali does not have a pH of 7.
c) Organic acids are neutralised slowly by bases.
d) The salt of the above acid and alkali is insoluble.

....

Q142: A buffer solution can be made from a:

a) strong acid and a salt of a weak acid.
b) weak acid and a salt of a strong acid.
c) weak acid and a salt of that acid.
d) strong acid and a salt of that acid.

....

Q143: 5 cm^3 of a 0.01 mol l^{-1} solution of hydrochloric acid was added to each of the following mixtures. The concentration of all the solutions is 0.1 mol l^{-1}.

In which case would there be the least change in pH?

a) 50 cm^3 NH_3 (aq) + 50 cm^3 HCl (aq)
b) 50 cm^3 NH_4Cl (aq) + 50 cm^3 NH_3 (aq)
c) 50 cm^3 HCl (aq) + 50 cm^3 Na Cl (aq)
d) 50 cm^3 NaCl (aq) + 50 cm^3 NH_4Cl (aq)

....

Q144: A buffer solution is made by dissolving 0.2 moles of sodium fluoride in one litre of hydrofluoric acid of concentration 0.1 mol l^{-1}.

Using information from the SQA data booklet, calculate the pH of the buffer solution.

....

Q145: A buffer solution of pH 4.6 was made up by dissolving sodium benzoate in a solution of benzoic acid.

If the concentration of the acid was 0.01 mol l^{-1}, calculate the concentration of the sodium benzoate, using information from the SQA data booklet.

....

When a pH electrode and meter are used to follow the titration between solutions of sodium hydroxide and methanoic acid, the pH graph shown below is obtained.

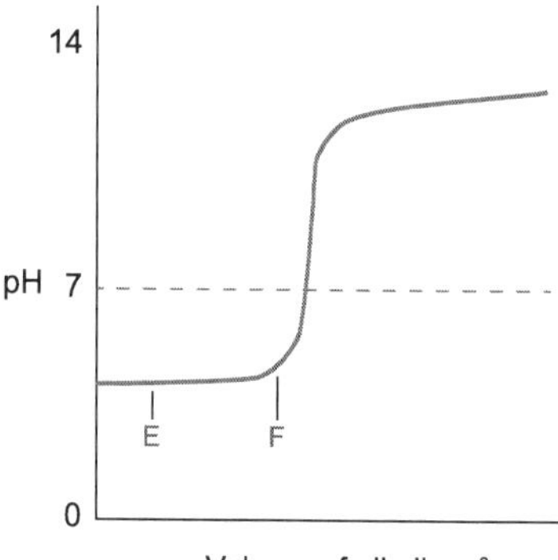

Q146: Which of the following indicators could be used to detect the endpoint?

a) Methyl orange indicator, pH range 3.0-4.4
b) Methyl red indicator, pH range 4.2-6.3
c) Phenol red indicator, pH range 6.8-8.4
d) Alizarin yellow R indicator, pH range 10.0-12.0

....

Q147: If the titration is stopped between E and F, the resulting solution acts as a buffer.

Why can this solution act as a buffer?

....

Q148: Explain how the pH of the buffer solution remains constant when a little more acid is added.

....

....

Topic 6

Reaction feasibility

Contents

Prerequisite knowledge

Before you begin this topic, you should know:

- *Standard Enthalpy of combustion - $\Delta H^o c$ is the energy released when one mole of a substance is burned completely in oxygen.*
- *Hess's Law - The energy change for a reaction is independent from the route taken to get from reactants to products.*
- *Bond Enthalpies - Energy required to break a bond and energy is given out when bonds are made.*
- *Thermochemistry is the study of energy changes that take place during chemical reactions.*

Learning objectives

By the end of this topic, you should understand:

- *how Ellingham diagrams can be used to predict the conditions under which a reaction can occur;*
- *that Ellingham diagrams can be used to predict the conditions required to* ***extract*** *a metal from its oxide;*
- *standard conditions;*

- *standard enthalpy of formation - definition and relevant calculations* $\Delta H^o = \Sigma H^o{}_f$ *(p) -* $\Sigma H^o{}_f$ *(r);*
- *entropy and prediction of the effect on entropy of changing the temperature or state;*
- *the second and third law of thermodynamics;*
- *standard entropy changes.* $\Delta S^o = \Sigma S^o$ *(p) -* ΣS^o *(r);*
- *the concept of free energy;*
- *the calculation of standard free energy change for a reaction* $\Delta G^o = \Sigma G^o$ *(p) -* ΣG^o *(r);*
- *applications of the concept of free energy;*
- *the prediction of the feasibility of a reaction under standard and non-standard conditions* ($\Delta G^o = \Delta H^o - T\Delta S^o$)*;*

6.1 Standard enthalpy of formation

The **standard state** of a substance is the most stable state of the substance under **standard conditions** and the standard conditions refer to a pressure of one atmosphere and a specific temperature, usually 298 K (25°C).

Key point

In calculations it is important to use temperature in Kelvin (K). To convert from ^{0}C to K simply add 273.

The standard **enthalpy of formation** of a compound $\Delta H^o{}_f$ is the energy given out or taken in when one mole of a compound is formed from its elements in their standard states.

For example the equation for the standard state of formation of ethanol is:

$$2C(s) + 3H_2(g) + {}^1/_2\, O_2(g) \rightarrow C_2H_5OH(l)$$

Standard enthalpy changes are measured in kJ mol^{-1} of reactant or product and the enthalpy change for a reaction is:

$$\Delta H^o = \Sigma\ \Delta H^o{}_{products} - \Sigma\ \Delta H^o{}_{reactants}$$

where Σ = mathematical symbol sigma for summation.

Enthalpy values	ΔH sign	Reaction type
ΣH(products) < ΣH(reactants)	Negative	Exothermic
ΣH(products) > ΣH(reactants)	Positive	Endothermic

6.2 Entropy

The **entropy** of a system is the measure of the disorder within that system; the larger the entropy the larger the disorder and vice versa. Entropy is given the symbol **S** and the standard entropy of a substance S^o is the entropy of one mole of the substance under standard conditions (1 atmosphere pressure and a temperature of 298K, 25°C). The units of entropy are J K^{-1} mol^{-1}.

Increasing entropy	Decreasing entropy
A puddle dries up on a warm day as the liquid becomes water vapour. The disorder (entropy) increases.	A builder uses a pile of loose bricks to construct a wall. The order of the system increases. Entropy falls.
Heating ammonium nitrate forms one mole of dinitrogen oxide and two moles of steam. Three moles of gas are formed. The disorder (entropy) increases.	When the individual ions in a crystal come together they take up a set position. The disorder of the system falls. Entropy decreases.
$NH_4NO_3(s) \rightarrow N_2O(g) + 2H_2O(g)$	$Na^+(aq) + Cl^-(aq) \rightarrow NaCl(s)$

Entropy values of substances in the solid state tend to be low due to the particles in solid occupying fixed positions. The particles are unable to move, but vibrate. Gases have high entropy values since their particles have complete freedom to move anywhere within the space they occupy. Entropy values for liquids lies somewhere between that of solids and gases.

Entropy and temperature

At 0K the particles in a solid no longer vibrate and are perfectly ordered. This means the entropy of a substance at 0K is zero (third law of Thermodynamics see next section). As the temperature increases, the entropy of the solid increases until the melting point, where there is a rapid increase in entropy as the solid melts into a liquid. A greater increase in entropy is seen when the liquid boils to become a gas.

Graph below shows how the entropy of a substance varies with temperature.

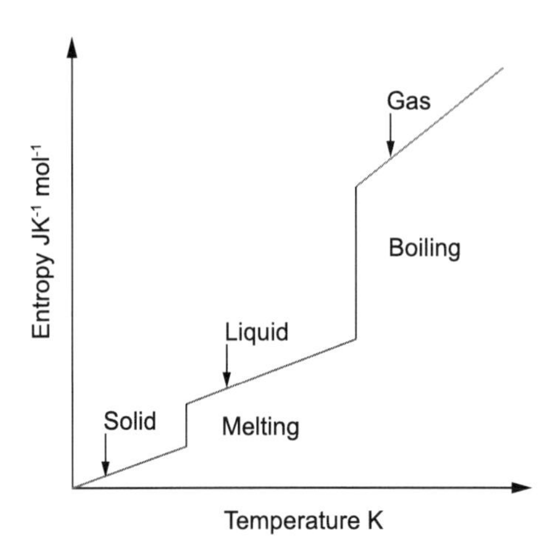

Calculating entropy in a chemical reaction

ΔS^o = standard entropy change in a chemical reaction can be worked out using:-

$$\Delta S^o = \Sigma \Delta S^o_{(products)} - \Sigma \Delta S^o_{(reactants)}$$

Q1: Using the data book and the following information calculate the standard entropy change for the following reaction.

$$2AgNO_3(s) \rightarrow 2Ag(s) + 2NO_2(g) + O_2(g)$$

S^o for $AgNO_3$ = 142 J K^{-1} mol^{-1}

S^o for NO_2 = 241 J K^{-1} mol^{-1}

Remember to multiply the entropy value by the number of moles of the chemical involved.

....

6.3 Second and third laws of thermodynamics

One version of the second law of thermodynamics defines the conditions of a feasible reaction. It states that for a reaction to be feasible, the total entropy change for a reaction system and its surroundings must be positive (the total entropy must increase).

$$\Delta S^{o}(\text{total}) = \Delta S^{o}(\text{system}) + \Delta S^{o}\ (\text{surroundings}) = +ve$$

Looking at the feasible reaction:

$$NH_3(g) + HCl(g) \rightarrow NH_4Cl(s)$$

ΔS^{o}(system) for this reaction is **-284** J K^{-1} mol^{-1}.

ΔH^{o} = -176 kJ mol^{-1} which means it is an exothermic reaction. The heat energy leaving the system causes the entropy of the surroundings to increase (hot surroundings have a higher entropy than cold surroundings).

ΔS^{o}(surroundings) = $-\Delta H^{o}/T$ where T is the temperature taken as the standard temperature 298K.

$$\text{So } \Delta S^{o}(\text{surroundings}) = -(-176)/298 = 0.591 \text{ kJ K}^{-1}\text{ mol}^{-1}\ (591 \text{ J K}^{-1}\text{ mol}^{-1})$$
$$\text{Therefore } \Delta S^{o}(\text{total}) = \Delta S^{o}(\text{system}) + \Delta S^{o}\ (\text{surroundings})$$
$$= -284 + 591 = +307 \text{ J K}^{-1}\text{ mol}^{-1}$$

The total entropy is positive confirming the reaction is feasible.

Q2: Calcium carbonate, present in limestone, is stable under normal atmospheric conditions. When it is in a volcanic area and it gets very hot, it can thermally decompose. Given the following information, use the Second Law of Thermodynamics to show why limestone is stable at 25°C, but not at 1500°C.

$CaCO_3(s) \rightarrow CaO(s) + CO_2(g)$ ΔH° = +178 kJ mol^{-1}

ΔS° = +161 J K^{-1} mol^{-1}

....

Q3: Graphite has been converted into diamond by the use of extreme pressure and temperature. Given the following information and values of entropy in the data booklet, show why diamond can **not** be made from graphite at 1 atmosphere pressure, either at room temperature or 5000°C.

$C(s)_{graphite} \rightarrow C(s)_{diamond}$ ΔH = +2.0 kJ mol^{-1}

....

Since the entropy of a substance depends on the order of the system, when a solid crystal is cooled to absolute zero (zero Kelvin), all the vibrational motion of the particles is stopped with each particle having a fixed location, i.e. it is 100% ordered. The entropy is therefore zero. This is one version of the 'third law of thermodynamics'.

As temperature is increased, entropy increases. As with enthalpy values, it is normal to quote standard entropy values for substances as the entropy value for the standard state of the substance.

Key point

Notice that the unit of entropy values is joules per kelvin per mole (J K^{-1} mol^{-1}). Be careful with this, since enthalpy values are normally in kilojoules per mole (kJ mol^{-1}) and in problems involving these quantities the units must be the same.

Go online

The second law of thermodynamics

An increase in entropy provides a driving force towards a reaction proceeding spontaneously. There are, however, processes that proceed spontaneously that seem to involve an entropy decrease.

For example, steam condenses to water at room temperature:

$$H_2O(g) \rightarrow H_2O(l) \qquad \Delta S^\circ = 70 - 189 = -119 \text{ J K}^{-1} \text{ mol}^{-1}$$
$$\Delta H^\circ = -44.1 \text{ kJ mol}^{-1}$$

In this case, the enthalpy change is also negative. This outpouring of energy from the system (- ΔH_{SYSTEM}) is transferred to the surroundings .

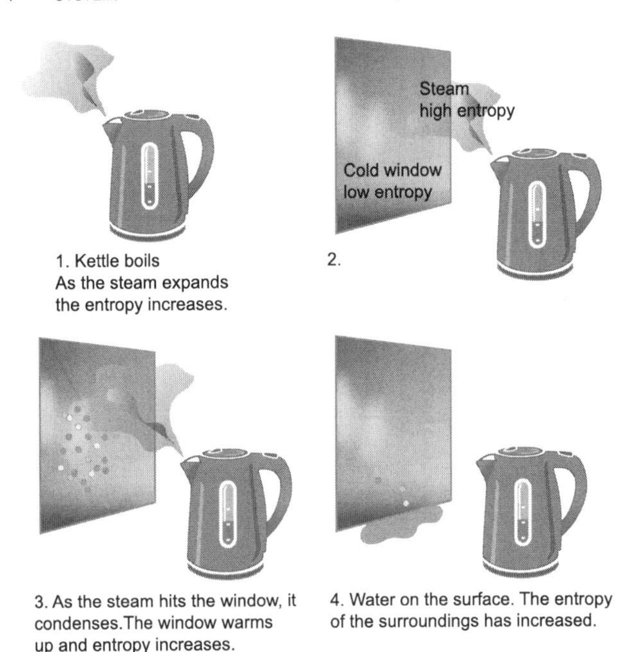

1. Kettle boils
As the steam expands the entropy increases.

2.

3. As the steam hits the window, it condenses.The window warms up and entropy increases.

4. Water on the surface. The entropy of the surroundings has increased.

The heat is transferred to the cold surface of the window and to the air around that area. This increases the disorder or entropy of the **surroundings**. (Just think of the scalding effect that would occur if your hand were placed in the steam - the disorder of the skin would increase!)

In general terms, heat energy released by a reaction system into the surroundings increases the entropy of the surroundings. If heat is absorbed by a reaction from the surroundings, this will decrease the entropy of the surroundings.

In the case of steam condensing, the entropy gain of the surroundings is equal to the energy lost (- ΔH) of the chemical system divided by the temperature:

$$\Delta S^o{}_{SURROUNDINGS} = \frac{-\ \Delta H^o{}_{SYSTEM}}{T}$$

The entropy change in the condensation situation therefore requires consideration of two entropy changes. The change in the system itself and the change in the surroundings must be added together, and for a spontaneous change to occur this total entropy change must be positive.

$$\Delta S^\circ_{TOTAL} = \Delta S^\circ_{SYSTEM} + \Delta S^\circ_{SURROUNDINGS}$$

In the case of steam condensing, the heat given out can be used to calculate a value for $\Delta S^\circ_{SURROUNDINGS}$:

$$\begin{aligned}
\Delta S^o{}_{SURR} &= \frac{-\ \Delta H_{SYSTEM}}{T} \\
\Delta S^o{}_{SURR} &= \frac{-\ (-\ 44.1\ \times\ 1000)}{298} \\
\Delta S^o{}_{SURR} &= +\ 148.0\ JK^{-1}mol^{-1}
\end{aligned}$$

And a calculation of ΔS°_{SYSTEM} from the data booklet gives:

All this means that although the entropy of a system itself may drop, the process itself will still be a natural, spontaneous change if the drop is compensated by a larger increase in entropy of the surroundings.

Expressed another way, this is the **second law of thermodynamics**. The total entropy of a reaction system and its surroundings always increases for a spontaneous change.

A word of caution: spontaneous does not mean 'fast'. It means 'able to occur without needing work to bring it about'. Thermodynamics is concerned with the **direction** of change and not the **rate** of change.

....

6.4 Free energy

We know that :-

ΔS^o(total) = ΔS^o(system) + ΔS^o (surroundings)

ΔS^o(surroundings) = $-\Delta H^o/T$

From this ΔS^o(total) = $-\Delta H^o/T$ + ΔS^o(system)

Multiplying this expression by -T we get $-T\Delta S^o$(total) = ΔH^o - $T\Delta S^o$(system)

- $T\Delta S^o$(total) has units of energy and we call this energy change the standard **free energy** change which is given the symbol ΔG^o.

$$\Delta G^o = \Delta H^o - T\Delta S^o$$

As ΔS^o(total) has to be positive for a reaction to be feasible ΔG^o must be negative.

The above equation can be used to predict whether a reaction is feasible or not.

$$NaHCO_3(s) \rightarrow Na_2CO_3(s) + CO_2(g) + H_2O(g)$$

The decomposition of sodium hydrogen carbonate:-

$\Delta H^o = +129 \text{ kJ mol}^{-1}$

$\Delta S^o = +335 \text{ J K}^{-1} \text{ mol}^{-1}$ ($0.335 \text{ kJ K}^{-1} \text{ mol}^{-1}$)

At 298K (standard temperature) $\Delta G^o = 129 - (298 \times 0.335) = +29 \text{ kJ mol}^{-1}$

ΔG^o at 298K is positive and therefore not feasible.

A reaction is feasible when ΔG^o is negative and therefore becomes feasible when ΔG^o = 0

$$0 = \Delta H^o - T\Delta S^o$$

$$\text{Therefore } T = \Delta H^o/\Delta S^o$$

For the above reaction the temperature at which it becomes feasible is 129/0.335 = 385K (112°C).

The standard free energy change of a reaction can be calculated from the standard free energies of formation of the products and reactants.

$$\Delta G^o = \Sigma\, \Delta G^o_{f(products)} - \Sigma\, \Delta G^o_{f(reactants)}$$

Go online

Calculations involving free energy changes

Q4: Use the table of standard free energies of formation to calculate values of ΔG° for these two reactions and thus predict whether or not the reaction is spontaneous .

a) $2Mg(s) + CO_2(g) \rightarrow 2MgO(s) + C(s)$
b) $2CuO(s) + C(s) \rightarrow 2Cu(s) + CO_2(g)$

SUBSTANCE	$\Delta G^\circ_{FORMATION}$ / kJ mol^{-1}
CO_2	-394
MgO	-569
ZnO	-318
CuO	-130
All elements	0

....

Q5: Calculate the standard free energy change at both 400 K and 1000 K for the reaction:

	$MgCO_3(s)$	$\rightarrow$	$MgO(s)$ +	$CO_2(g)$
ΔH°_f /kJ mol^{-1}	-1113		-602	-394
S°/J K^{-1} mol^{-1}	66		27	214

....

Q6: Use the data given, along with data booklet values to calculate the temperature at which the Haber process becomes feasible.

	$N_2(g)$ +	$3H_2(g)$	$\rightleftharpoons$	$2NH_3(g)$
ΔH°_f /kJ mol^{-1}	0	0		-46.4
S°/J K^{-1} mol^{-1}	?	?		193.2

....

Q7: Given these reaction values for oxides of nitrogen:

$4NO(g) \rightarrow 2N_2O(g) + O_2(g)$ $\Delta G^\circ = -139.56$ kJ

$2NO(g) + O_2(g) \rightarrow 2NO_2(g)$ $\Delta G^\circ = -69.70$ kJ

a) Calculate ΔG° for this reaction:
$2N_2O(g) + 3O_2(g) \rightleftharpoons 4NO_2(g)$

b) Say whether the equilibrium position favours reactants or products.

....

Q8: Chloroform was one of the first anaesthetics used in surgery. At the boiling point of any liquid, the gas and liquid are in equilibrium. Use this information to calculate a boiling point for chloroform.

$CHCl_3(l) \rightarrow CHCl_3(g)$ $\Delta H^\circ = 31.4$ kJ mol^{-1}

$\Delta S^\circ = 94.2$ J K^{-1} mol^{-1}

....

....

Free energy and equilibrium

ΔG^o for a reaction can give information about the equilibrium position in a reversible reaction and the value of the equilibrium constant K.

$\Delta G^o < 0$ the forward reaction will be feasible and therefore the products will be favoured over the reactants. The equilibrium position will lie to the right side (products) of the equilibrium and K will be greater than 1.

$\Delta G^o > 0$ the backwards reaction will be feasible and therefore the reactants will be favoured over the products. The equilibrium position will lie to the left side (reactants) of the equilibrium and K will be less than 1.

Equilibrium reaction $R \leftrightharpoons P$

If ΔG^o is negative and we start with 1 mole of pure R at 1 atmosphere of pressure standard state conditions apply and so at the start of the reaction we can talk about the standard free energy of R as opposed to the free energy of R.

As soon as the reaction starts and some R is converted to P standard state conditions no longer apply and therefore during a chemical reaction we talk about the free energy rather than the standard free energy. As we approach equilibrium the free energy moves towards a minimum. As ΔG^o is negative the products are favoured and the equilibrium lies to the right (products).

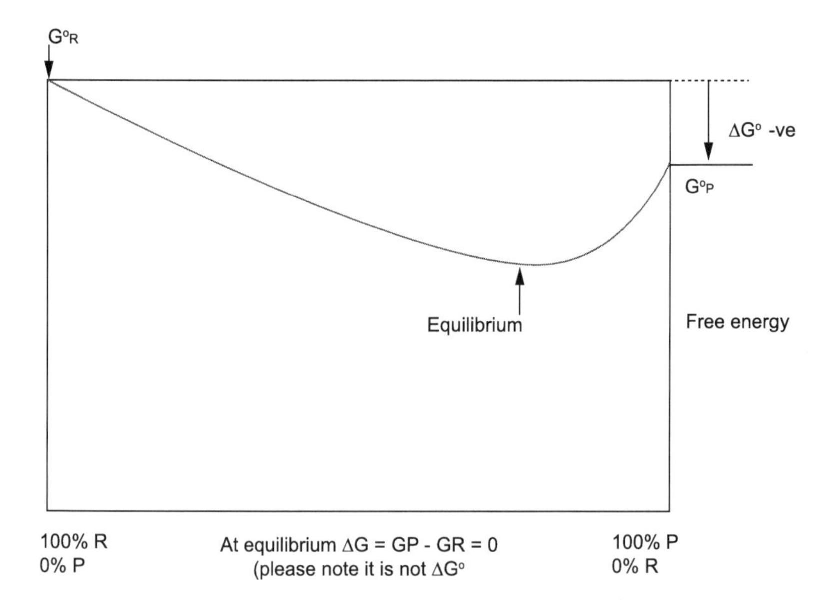

6.5 Ellingham diagrams

$\Delta G^o = \Delta H^o - T\Delta S^o$ can be rearranged to $\Delta G^o = -\Delta S^o T + \Delta H^o$.

Comparing this with the equation for a straight line y =mx+ c we can see from a plot of free energy change against temperature will have a gradient of $-\Delta S^o$ and an intercept on the y-axis of ΔH^o. This is known as an Ellingham diagram.

Plotting an Ellingham diagram

This table shows values of ΔG° over a temperature range for the formation of water gas (Equation 6.1) .

Temperature K	200	400	600	800	1000	1200
ΔG° / kJ mol^{-1}	104	78	51	24	-3	-29

This graph shows values of ΔG° over a temperature range for the formation of water gas referring to:

$$C(s) + H_2O(g) \rightarrow CO(g) + H_2(g) \tag{6.1}$$

....

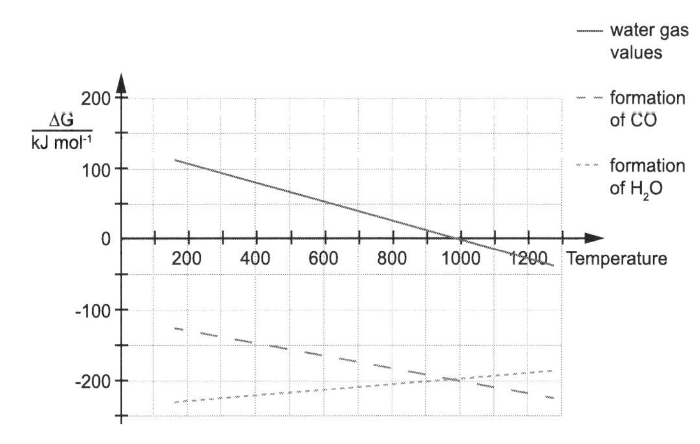

This reaction can be considered as a combination of these two oxide formations:

$$C(s) + {}^1/_2O_2(g) \rightarrow CO(g) \tag{6.2}$$

....

$$H_2(g) + {}^1/_2O_2(g) \rightarrow H_2O(g) \tag{6.3}$$

....

If the second equation (Equation 6.3) was reversed and added to the first (Equation 6.2) then the formation of water gas (Equation 6.1) results.

Both Equation 6.2 and Equation 6.3 have their values for ΔG° at various temperatures calculated and plotted onto the same graph.

Formation of CO(g)

Temperature K	200	400	600	800	1000	1200
ΔG° / kJ mol^{-1}	-128	-146	-164	-182	-200	-218

Formation of $H_2O(g)$

Temperature K	200	400	600	800	1000	1200
ΔG° / kJ mol^{-1}	-223	-224	-215	-206	-197	-188

Interpretation

The lines relating ΔG to temperature for the two equations Equation 6.2 and Equation 6.3 intersect at 981.3 K when $\Delta G = 0$. The carbon and hydrogen are both capable of reacting with oxygen but at any temperature **above** 981.3 K, the carbon is capable of winning oxygen from the water molecule and **forcing** the second equation (Equation 6.3) to reverse. **Consider 1000 K**

$C(s) + {}^{1}/{}_{2}O_2(g) \rightarrow CO(g)$ $\quad \Delta G^\circ = -200$ kJ mol^{-1}

$H_2(g) + {}^{1}/{}_{2}O_2(g) \rightarrow H_2O(g)$ $\quad \Delta G^\circ = -197$ kJ mol^{-1}

By reversing the second equation and adding to the first, the result is:

$C(s) + H_2O(g) \rightarrow CO(g) + H_2(g)$ $\quad \Delta G^\circ = -3$ kJ mol^{-1}

At 1000 K the formation of water gas is feasible and spontaneous.

Consider 800 K

$C(s) + {}^{1}/{}_{2}O_2(g) \rightarrow CO(g)$ $\quad \Delta G^\circ = -182$ kJ mol^{-1}

$H_2(g) + {}^{1}/{}_{2}O_2(g) \rightarrow H_2O(g)$ $\quad \Delta G^\circ = -206$ kJ mol^{-1}

By reversing the second equation and adding to the first, the result is:

$C(s) + H_2O(g) \rightarrow CO(g) + H_2(g)$ $\quad \Delta G^\circ = +24$ kJ mol^{-1}

At 800 K the formation of water gas is **not** thermodynamically feasible or spontaneous. The Ellingham diagram provides a simple clear picture of the relationship between the different reactions and allows prediction of the conditions under which combinations of individual reactions become feasible.

....

6.5.1 Extraction of metals

Ellingham diagrams plot values of ΔG° against temperature. If the lines are drawn for metal oxide formation reactions, these can be used to predict the conditions required to **extract** a metal from its oxide. This requires the formation of the metal oxide process to be **reversed**. Any chemical used to aid the reversing of this process must provide enough free energy to supply this reversal. It is normal to write all reactions that are on the graph to involve one mole of oxygen (so that oxygen is removed when two equations are combined).

Interpreting Ellingham diagrams

Answer these questions on paper before displaying the explanation. In each case refer to the Ellingham diagram below.

Figure 6.1: Ellingham diagram

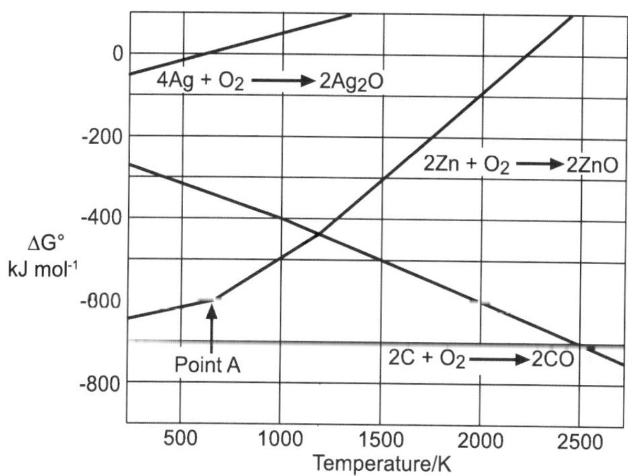

. . . .

Q9: Which oxide (in Figure 6.1) could be broken down by heat alone at 1000 K? (Hint: at 1000 K the ΔG value of the **reversed** reaction needs to be negative.)

. . . .

Q10: Above which temperature would the breakdown of zinc oxide become feasible by heat alone?

. . . .

Q11: Use the graph Figure 6.1 to calculate the ΔG° value for the reaction in which carbon reduces zinc oxide at:

a) 1000 K
b) 1500 K

$$2C(s) + 2ZnO(s) \rightarrow 2Zn(s) + 2CO(g)$$

Try each calculation for a) and b) on paper by following this route.

i. Write down the target equation.
ii. Write the equations for carbon combustion and zinc combustion, along with their ΔG° values from the graph at 1000 K in Figure 6.1.
iii. Write the reversed equation for the zinc combustion remembering to reverse ΔG°.
iv. Add this new equation to the carbon equation and note the sign on ΔG°. Is the reaction feasible or not?

....

Q12: At what temperature does the reduction of zinc oxide by carbon become feasible?

....

Q13: The data booklet gives the melting point of zinc as approximately 700 K. What happens to the entropy and what effect does it have on the gradient of the graph? (point A)

....

Q14: What causes the further little 'kink' in the zinc line at 1180 K?

....

....

Try the next two questions yourself, on paper.

Q15: This Ellingham diagram below shows the reactions involved in the blast furnace reduction of iron(II) oxide with carbon.

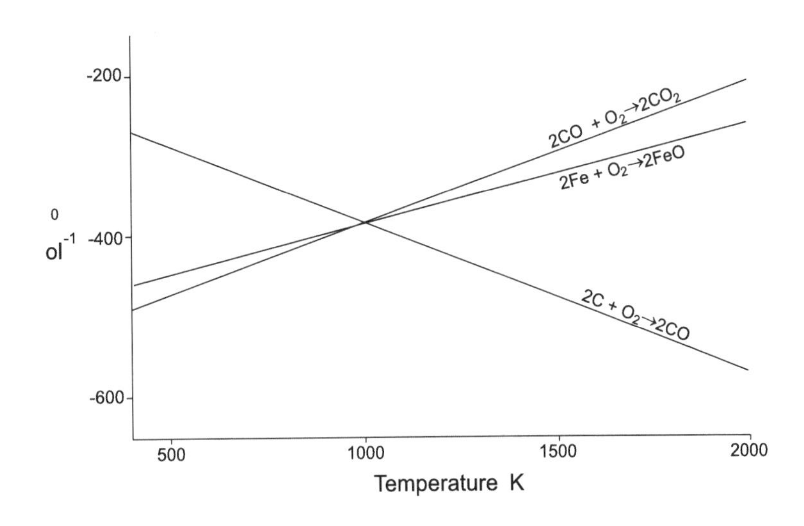

a) Write the combined equation showing the reduction of iron(II) oxide by carbon at 1500 K.

b) Calculate the standard free energy change at this temperature.

c) At what temperature does the reduction of iron(II) oxide by carbon become feasible?

d) At what temperatures will it be thermodynamically feasible for carbon monoxide to reduce iron(II) oxide?

e) Can you suggest a reason (apart from the temperature) why carbon monoxide might be more efficient than carbon at reducing iron(II) oxide?

....

Q16: Although magnesium ores are very abundant in the Earth's crust, the very high reactivity of magnesium makes it difficult to extract the metal. During the Second World War, magnesium was manufactured by reduction of its oxide by carbon.

$$2MgO + 2C \rightarrow 2CO + 2Mg$$

Examine the Ellingham diagram, Figure 6.2, and answer the questions which follow.

Figure 6.2: Ellingham diagram

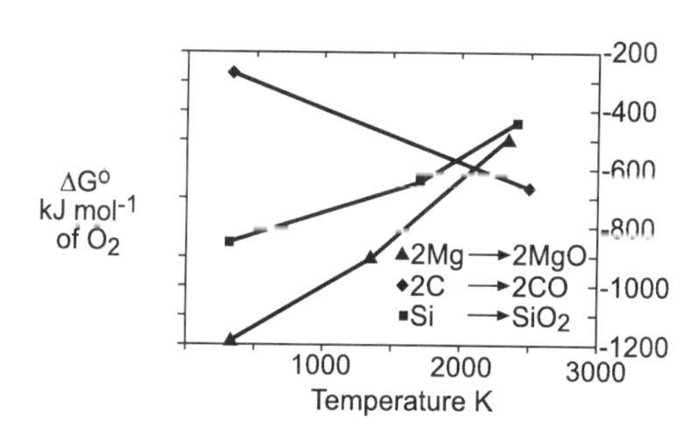

....

a) In what temperature range is the above process thermodynamically feasible?

b) Describe **two** problems that the operation of the process at this temperature would present.

c) Use the Ellingham diagram to calculate ΔG° for the production of magnesium in the following:

$2MgO + Si \rightarrow SiO_2 + 2Mg$ at 1500K

In industry, the extraction of magnesium from magnesium oxide using silicon involves two modifications.

i. A mixture of calcium oxide and magnesium oxide is used, and the calcium oxide reacts with the silicon oxide produced.

$$CaO + SiO_2 \rightarrow CaSiO_3$$
$$(\Delta G^\circ = -92 \text{ kJ mol}^{-1} \text{ at } 1500 \text{ K})$$

ii. The gaseous magnesium formed is continuously removed from the reaction mixture.
iii. Use this information to answer the questions below.

d) Calculate ΔG° for the reaction:
$CaO + 2MgO + Si \rightarrow CaSiO_3 + 2Mg$ at 1500 K

e) Explain why the removal of magnesium from the reaction mixture helps the process.

....

6.6 Summary

Summary

You should now understand:

- how Ellingham diagrams can be used to predict the conditions under which a reaction can occur;
- that Ellingham diagrams can be used to predict the conditions required to **extract** a metal from its oxide;
- standard conditions;
- standard enthalpy of formation - definition and relevant calculations $\Delta H^o = \Sigma H^o{}_f\ (p) - \Sigma H^o{}_f\ (r)$;
- entropy and prediction of the effect on entropy of changing the temperature or state;
- the second and third law of thermodynamics;
- standard entropy changes. $\Delta S^o = \Sigma S^o\ (p) - \Sigma S^o\ (r)$;
- the concept of free energy;
- the calculation of standard free energy change for a reaction $\Delta G^o = \Sigma G^o\ (p) - \Sigma G^o\ (r)$;
- applications of the concept of free energy;
- the prediction of the feasibility of a reaction under standard and non-standard conditions ($\Delta G^o = \Delta H^o - T\Delta S^o$).

6.7 Resources

- Department of Physics University of Toronto: Einstein Quote (http://bit.ly/29TG6Wr)
- BBC: The Second Law of Thermodynamics (http://bbc.in/29VpLMZ)
- ENDOTHERMIC reactions (http://bit.ly/29QsGos)
- ChemConnections (http://bit.ly/29QsGos)
- Former AH Chemistry Unit 2 PPA 4, Verification of a Thermodynamic Prediction (http://bit.ly/2apLIFy) .

6.8 End of topic test

Go online

End of Topic 6 test

The end of topic test for *Reaction feasibility.*

Q17: Which of the following reactions results in a large **decrease** in entropy?

a) $CaCO_3(s) \rightarrow CaO(s) + CO_2(g)$
b) $C(s) + H_2O(g) \rightarrow CO(g) + H_2(g)$
c) $N_2O4(g) \rightarrow 2NO_2(g)$
d) $N_2(g) + 3H_2(g) \rightarrow 2NH_3(g)$

....

Q18: At 1400 K
$2C + O_2 \rightarrow 2CO$ ΔG° = -475 kJ mol^{-1} of O_2
$2Zn + O_2 \rightarrow 2ZnO$ ΔG° = -340 kJ mol^{-1} of O_2

For the reaction
$C + ZnO \rightarrow Zn + CO$

the standard free energy change at 1400 K is

a) 135 kJ mol^{-1} of ZnO
b) 67.5 kJ mol^{-1} of ZnO
c) -67.5 kJ mol^{-1} of ZnO
d) -135 kJ mol^{-1} of ZnO

....

Q19: The standard entropy values (J K^{-1} mol^{-1}) for a number of compounds are shown below.

Compound	$CH_{4(g)}$	$O_{2(g)}$	$CO_{2(g)}$	$H_2O_{(g)}$	$H_2O_{(l)}$
S°	186	205	214	189	70

The standard entropy change (J K^{-1} mol^{-1}) for the complete combustion of one mole of methane is:

a) 242
b) -37
c) -242
d) -4

....

Q20: ΔG° gives an indication of the position of the equilibrium for a reaction.

The equilibrium lies on the side of the products when ΔG° is:

a) zero.
b) large and positive.
c) large and negative.
d) one.

....

Q21: At Tb, the boiling point of a liquid

$$\Delta S_{vaporisation} = \frac{\Delta H_{vaporisation}}{T_b}$$

For many liquids,
$\Delta S_{vaporisation}$=88 Jk^1 mol^{-1} (approx)

If this value was true for water ($\Delta H^{\circ}_{vaporisation}$ = 40.6 kJ mol^{-1}), the predicted boiling point of water would be:

a) 0.46 K
b) 373 K
c) 461 K
d) 2.16 K

....

Q22:

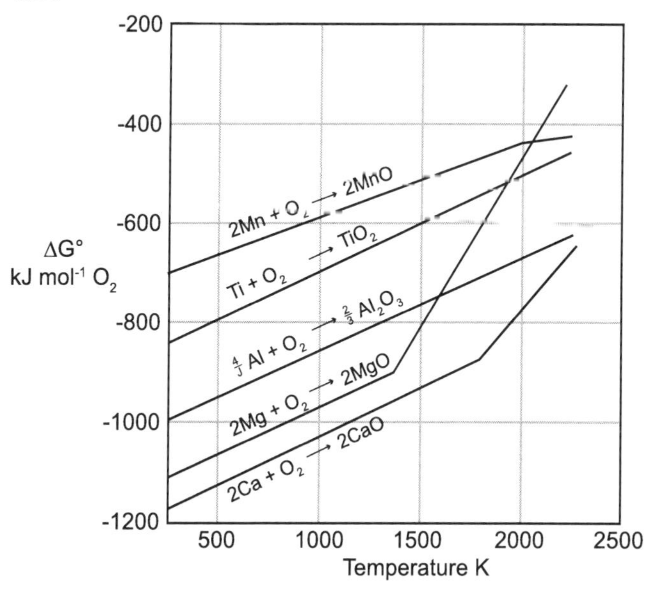

The reduction of Al_2O_3 to Al is thermodynamically feasible at 2000 K using:

a) titanium.
b) calcium.
c) magnesium.
d) manganese.

....

Consider the following reactions and their values for ΔG° and ΔH° at 298 K.

Reaction	**ΔG° (kJ mol^{-1})**	**ΔH° (kJ mol^{-1})**
$\frac{1}{2} H_2(g) + \frac{1}{2} Cl_2(g) \rightarrow HCl(g)$	-95	-92
$2 Al(s) + 1\frac{1}{2} O_2(g) \rightarrow Al_2O_3(s)$	-1576	-1669
$NH_4Cl(s) \rightarrow NH_4^+(aq) + Cl$-(aq)	-7	16

Q23: The second reaction has the greatest difference in the values of ΔG° and ΔH°. Suggest a reason for this difference.

....

Q24: From the values given for the third reaction, it can be concluded that ammonium chloride dissolves **spontaneously** in water under standard conditions with a drop in **temperature**.
Why can we come to these two conclusions?

....

Q25: Calculate the entropy change for the first reaction at 298 K.

....

Q26: Benzene boils at 80°C. The entropy change at this temperature is shown.
Calculate a value for the enthalpy change on boiling (the enthalpy of vaporisation).
$C_6H_6(l) \rightleftharpoons C_6H_6(g)$ $\Delta S^\circ_{vaporisation} = 97.2$ J K^{-1} mol^{-1}
Give your answer in kJ mol^{-1} to one decimal place and with a sign.

....

....

Topic 7

Kinetics

Contents

Learning objectives

By the end of this topic, you should know:

- *how to determine the order of a reaction from experimental data and rate equations;*
- *how to calculate the rate constant and its units;*
- *how to predict from the rate equation the rate determining step and a possible mechanism.*

7.1 Determination of order of reaction

Kinetics is about how fast a reaction goes (the rate of the chemical reaction).

The rate of a chemical reaction normally depends on the concentrations of the reactants.

A + B $\rightarrow$ Products

If we double the initial concentration of A and keep the initial concentration of B constant the rate of reaction doubles. This suggests that the rate of reaction is directly proportional to the concentration of A so **rate α $[A]^1$**.

If the rate increases by a factor of four when the initial concentration of B is doubled and the initial concentration of A is kept constant it suggests that the rate is directly proportional to the square of the concentration of B so **rate α $[B]^2$.**

If we combine these results it gives us **rate α $[A]^1[B]^2$**.

This can be re-written as **rate = k $[A]^1[B]^2$** where k is the rate constant. This reaction would be **first** order with respect to A and **second** order with respect to B.

For a reaction of the type aA + bB $\rightarrow$ products we can express how the rate depends on the concentrations of A and B using the following expression **rate = k $[A]^m[B]^n$**. The indices m and n are the orders of the reaction with respect to A and B respectively (they bear no resemblance to the stoichiometric coefficients in balancing the chemical equation). They are usually small whole numbers no greater than 2. The overall order of the reaction is given by m + n so in the above reaction the order would be 1 + 2 = 3, the reaction would be **third** order overall.

The rate constant k units depend on the overall order of the reaction.

Overall order	**Units of k**
0	mol l^{-1} s^{-1}
1	s^{-1}
2	mol^{-1} l s^{-1} (or l mol^{-1} s^{-1})
3	mol^{-2} l^2 s^{-1} (or l^2 mol^{-2} s^{-1})

Example:

rate = k $[A]^2[B]^1$

k = rate/$[A]^2[B]^1$

Rate measured in mol l^{-1} s^{-1} and concentration in mol l^{-1}

k = mol l^{-1} s^{-1}/ (mol $l^{-1})^2$ (mol l^{-1})

k = mol l^{-1} s^{-1}/ (mol^2 l^{-2}) (mol l^{-1})

k = mol^{-2} l^2 s^{-1}

7.2 Calculation of rate constants

The rate equation for a chemical reaction can only be determined experimentally. This is done through a series of experiments where the initial concentrations are varied. The initial rate for each experiment is determined.

A + B + C → Products

Experiment	[A] mol l^{-1}	[B] mol l^{-1}	[C] mol l^{-1}	Initial rate mol l^{-1} s^{-1}
1	1.0	1.0	1.0	20
2	2.0	1.0	1.0	40
3	1.0	2.0	1.0	20
4	1.0	1.0	2.0	80

If we compare experiments 1 and 2 we can see that doubling the initial concentration of A causes the rate to increase by a factor of 2. This implies the reaction is first order with respect to A.

Comparing reactions 1 and 3 we can see that doubling the initial concentration of B has no effect on the initial rate of reaction implying that the reaction is zero order with respect to B.

Comparing reactions 1 and 4 we can see that doubling the initial concentration of C causes the initial rate of reaction to increase by a factor of 4. This implies the reaction is second order with respect to C.

The rate equation for this reaction would be rate = k $[A]^1[B]^0[C]^2$ more simply written as

Rate = k $[A]^1[C]^2$

The reaction is third order overall and the rate constant would have units of $mol^{-2}\ l^2\ s^{-1}$.

To calculate the rate constant k we can use any one of the above four reactions.

$k = rate/[A]^1[C]^2 = 20/1.0 \times (1.0)^2 = 20/1.0 = 20\ mol^{-2}\ l^2\ s^{-1}$

Orders and rate constants

Go online

The next two questions refer to the following reaction:

Bromide ions are oxidised by bromate ions (BrO_3^-) in acidic solution according to the equation:

$$5Br^-(aq) + BrO_3^-(aq) + 6H^+(aq) \rightarrow 3Br_2(aq) + 3H_2O(\ell)$$

By experiment, the reaction is found to be first order with respect to both bromide and bromate but second order with respect to hydrogen ions.

Q1: Write the rate equation for this reaction.

....

Q2: What is the overall order of the reaction?

....

The next four questions refer to the hydrolysis of urea in the presence of the enzyme, urease.

$$NH_2CONH_2(aq) + H_2O(l) \rightarrow 2NH_3(g) + CO_2(g)$$

The rate equation for the reaction is found by experiment to be:

$$\text{Rate} = k\,[\text{urea}][\text{urease}]$$

Q3: What is the overall order of reaction?

. . . .

Q4: What is the order with respect to water?

. . . .

Q5: What is the order with respect to urea?

. . . .

Q6: What is the order with respect to urease?

. . . .

The next four questions refer to the decomposition of dinitrogen pentoxide, N_2O_5:

$$2N_2O_5(g) \rightarrow 4NO_2(g) + O_2(g)$$

Experiments were carried out in which the initial concentration was changed and the initial rate of reaction was measured. The following data were obtained.

$[N_2O_5]$ / mol ℓ^{-1}	**Initial Rate / mol ℓ^{-1} s^{-1}**
0.05	2.2×10^{-5}
0.10	4.4×10^{-5}
0.20	8.8×10^{-5}

Q7: Write the rate equation for the reaction.

. . . .

Q8: What will be the units of the rate constant?

. . . .

Q9: Calculate the rate constant for the reaction. (Do not include units.)

. . . .

Q10: If the initial concentration was 0.07 mol ℓ^{-1}, calculate the initial rate of reaction in mol ℓ^{-1} s^{-1}. (Do not include units.)

. . . .

The next three questions refer to the following reaction:

Iodide ions are oxidised in acidic solution to triiodide ions, I_3^-, by hydrogen peroxide.

$$H_2O_2(aq) + 3I^-(aq) + 2H^+(aq) \rightarrow I_3^-(aq) + 2H_2O(\ell)$$

The following initial rate data were obtained:

Experiment	Initial Concentrations / mol ℓ^{-1}			Initial Rate / mol ℓ^{-1} s^{-1}
	$[H_2O_2]$	$[I^-]$	$[H^+]$	
1	0.02	0.02	0.001	9.2×10^{-6}
2	0.04	0.02	0.001	1.84×10^{-5}
3	0.02	0.04	0.001	1.84×10^{-5}
4	0.02	0.02	0.002	9.2×10^{-6}

Q11: From the above data, write the rate equation.

....

Q12: What will be the units of k?

....

Q13: Calculate the value for the rate constant.

....

....

7.3 Reaction mechanisms

Chemical reactions usually happen by a series of steps rather than by one single step. This series of steps is known as the reaction mechanism. The overall rate of a reaction is dependent on the slowest step, which is called the rate-determining step.

$$2NO_2 + F_2 \rightarrow 2NO_2F$$

$$\text{Rate} = k\,[NO_2][F_2]$$

The reaction is first order with respect to each of the reactants and this suggests that 1 molecule of each of the reactants must be involved in the slow **rate determining step**.

$$NO_2 + F_2 \rightarrow NO_2F + F \text{ (slow step)}$$

$$NO_2 + F \rightarrow NO_2F \text{ (fast step)}$$

Adding the two steps together gives the overall equation for the reaction.

Please note that an experimentally determined rate equation can provide evidence but not proof for a proposed reaction mechanism.

Do we need to know how fast each step is in order to work out the overall rate? As an analogy, consider this production line in the bottling plant in a distillery.

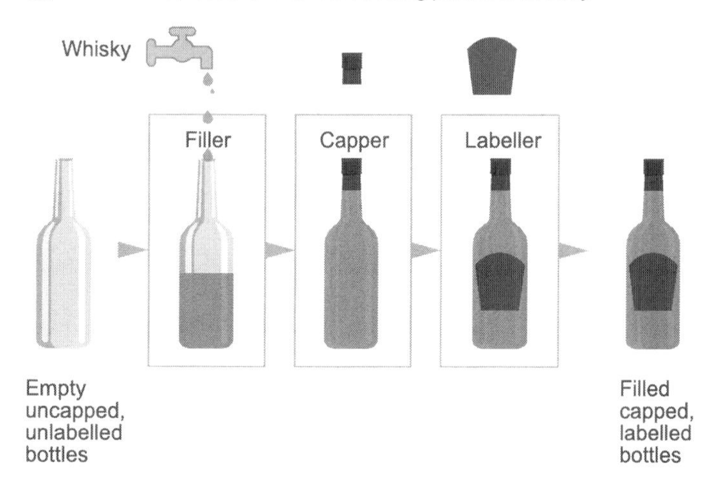

A whisky production line

There are three parts and the maximum capacity of each is:

Part 1	Filler	2 bottles filled per minute.
Part 2	Capper	120 bottles capped per minute.
Part 3	Labeller	60 bottles labelled per minute.

The production line is switched on. After 30 seconds a bottle has been filled and is passed to the Capper which caps it in 0.5 seconds and passes it to the Labeller which takes a further second to label it. So after 31.5 seconds we have completed one bottle. After one minute, the second bottle is full and is immediately capped by the Capper, which has been twiddling its thumbs waiting for the bottle to arrive.

Q14: How many seconds will it take to fill, cap and label 100 bottles?

....

Q15: How many seconds will it take to fill, cap and label 1000 bottles?

....

Clearly it does not matter how fast the Capper and Labeller are. The overall process is governed by how fast the bottles are filled. Bottle filling is the rate determining step.

In any chemical reaction mechanism, one step will be significantly slower than the others and this step will determine the overall reaction rate, i.e. it will be the rate determining step (RDS for short).

Key point

In general, the overall rate of a reaction depends on the rate of the slowest step in the mechanism. The rate equation provides information about the rate-determining step.

Questions on reaction mechanisms

Q16: Which of the following reactions is **most** likely to occur by a simple one-step process?

a) $4HBr + O_2 \rightarrow 2H_2O + 2Br_2$
b) $H_2S + Cl_2 \rightarrow S + 2HCl$
c) $2NO + O_2 \rightarrow 2NO_2$
d) $2H_2 + O_2 \rightarrow 2H_2O$

....

Q17: Which of the following reactions is **least** likely to occur by a simple one-step process?

a) $4HBr + O_2 \rightarrow 2H_2O + 2Br_2$
b) $H_2S + Cl_2 \rightarrow S + 2HCl$
c) $2NO + O_2 \rightarrow 2NO_2$
d) $2H_2 + O_2 \rightarrow 2H_2O$

....

The next six questions refer to the reaction between propanone and bromine in alkaline solution.

The balanced equation is:

$$CH_3COCH_3(aq) + Br_2(aq) + OH^-(aq) \rightarrow CH_3COCH_2Br(aq) + H_2O(\ell) + Br^-(aq)$$

The experimentally determined rate equation is:

$$\text{Rate} = k\,[CH_3COCH_3][OH^-]$$

Use this information to select True or False for each of the following statements.

Q18: The reaction is first order with respect to bromine.

a) True
b) False

....

Q19: The reaction involves a simple one-step process.

a) True
b) False

....

Q20: The reaction is second order overall.

a) True
b) False

....

Q21: The rate determining step involves one molecule of propanone and one molecule of bromine.

a) True
b) False

....

Q22: The following mechanism fits the rate equation.

$$CH_3COCH_3 \longrightarrow CH_3C(OH){=}CH_2 \quad \textbf{slow}$$

$$CH_3C(OH){=}CH_2 \longrightarrow CH_3C(OH)(Br)CH_2Br \quad \textbf{fast}$$

$$CH_3C(OH)(Br)CH_2Br + {}^{\ominus}OH \longrightarrow CH_3COCH_2Br + H_2O + Br^{\ominus} \quad \textbf{fast}$$

a) True
b) False

....

Q23: The following mechanism fits the rate equation.

$$CH_3COCH_3 + {}^{\ominus}OH \longrightarrow CH_3CO\overset{\ominus}{C}H_2 + H_2O \quad \textbf{slow}$$

$$CH_3CO\overset{\ominus}{C}H_2 + Br_2 \longrightarrow CH_3COCH_2Br + Br^{\ominus} \quad \textbf{fast}$$

a) True
b) False

....

The following questions refer to a reaction involving hydrogen peroxide and bromide ions in aqueous solution.

$H_2O_2 + Br^-$	$\rightarrow$	$BrO^- + H_2O$	Step 1
$H_2O_2 + BrO^-$	$\rightarrow$	$Br^- + H_2O + O_2$	Step 2

Q24: What is the equation for the overall reaction ?

....

Q25: What is the role played by the Br^- ion?

....

Q26: What role is played by the BrO^- ion?

....

Q27: If step 1 is the rate determining step, which of the following is the rate equation?

a) Rate = k $[H_2O_2]$
b) Rate = k $[H_2O_2][Br^-]$
c) Rate = k $[H_2O_2]^2$
d) Rate = k $[H_2O_2]^2$ $[Br^-]$

....

....

7.4 Summary

Summary

You should now be able to:

- determine the order of a reaction from experimental data and rate equations;
- calculate the rate constant and its units;
- predict from the rate equation the rate determining step and a possible mechanism.

7.5 Resources

- SSERC Bulletin (page 9) (http://bit.ly/29NFDnF)
- Finding orders of reaction (http://bit.ly/29QxVok)
- Royal Society of Chemistry: Chemistry at the Races (http://rsc.li/29WOApS)

7.6 End of topic test

End of Topic 7 test

Go online

The end of topic test for *Kinetics.*

Q28: Which of the following is the unit for the rate of a chemical reaction?

a) mol l^{-1}
b) s mol^{-1}
c) mol l^{-1} s^{-1}
d) s^{-1}

....

The following data refer to initial reaction rates obtained for the reaction:

X + Y + Z → products

Experiment	Rate concentration			Relative initial rate
	[X]	[Y]	[Z]	
1	1.0	1.0	1.0	0.3
2	1.0	2.0	1.0	0.6
3	2.0	2.0	1.0	1.2
4	2.0	1.0	2.0	0.6

Q29: The following data refer to initial reaction rates obtained for the reaction:

a) Rate = k [X] [Y]
b) Rate = k [X] [Y] [Z]
c) Rate = k [X] $[Y]^2$
d) Rate = k [X]

....

Q30: The rate of a particular chemical reaction is first order with respect to each of two reactants. The units of k, the rate constant, for the reaction are:

a) mol^{-1} l s^{-1}
b) l^2 mol^{-2} s^{-2}
c) mol l^{-1} s^{-1}
d) mol^2 l^{-2} s^{-2}

....

Q31: $2X + 3Y_2 \rightarrow$ products
A correct statement which can be made about the above reaction is that:

a) the reaction will be slow due to the number of particles colliding.
b) the reaction order with respect to X is 2.
c) the overall order of the reaction is 5.
d) the rate expression cannot be predicted.

....

Q32: $P + Q \rightarrow R$
The rate equation for this reaction is:
Rate = k [P] $[Q]^2$
If the concentration of P and Q are both doubled, how many times will the rate increase?

a) 2
b) 4
c) 6
d) 8

....

Q33: The reaction expressed by the stoichiometric equation
$Q + R \rightarrow X + Z$
was found to be first order with respect to each of the two reactants.
Which of the following statements is correct?

a) The rate of the reaction is independent of either Q or R.
b) The rate of reaction decreases as the reaction proceeds.
c) Overall, the reaction is first order.
d) If the initial concentrations of both Q and R are halved, the rate of reaction will be halved.

....

Q34: The reaction:
$X + 2Y \rightarrow Z$
has a rate equation of the form:
Rate = k[X][Y]
If the reaction proceeds by a two step process, then the rate-determining step might be:

a) **$X + Y \rightarrow Z$**
b) **$X + 2Y \rightarrow$ intermediate**
c) **$X + Y \rightarrow$ intermediate**
d) **$XY + Y \rightarrow Z$**

....

Q35: For the reaction
$NO(g) + N_2O_5(g) \rightarrow 3NO_2(g)$
the following mechanism is suggested.
Step 1: $N_2O_5(g) \rightarrow NO_2 + NO_3(g)$ **slow**
Step 2: $NO(g) + NO_3(g) \rightarrow 2NO_2(g)$ **fast**
Experimental evidence to support this would be obtained if the rate of the reaction equals:

a) k[NO]
b) $k[N_2O_5][NO]$
c) $k[NO][NO_3]$
d) $k[N_2O_5]$

....

Q36: For a given chemical change involving two reactants P and Q, rate of reaction is directly proportional to [P][Q].
If the equation representing the overall reaction is
P + 2Q → S + T
the mechanism could be:

a) **2Q → R + S fast**
R + P → T slow
b) **P + Q → R + S fast**
R + Q → T slow
c) **P + Q → R + S slow**
R + Q → T fast
d) **P → R + S fast**
R + 2Q → T slow

....

Q37: A reaction
A + B → C + D
is found to be first order with respect to B and zero order with respect to A.
Which of the following graphs is consistent with these results? Answer by clicking on the appropriate graph.

a)

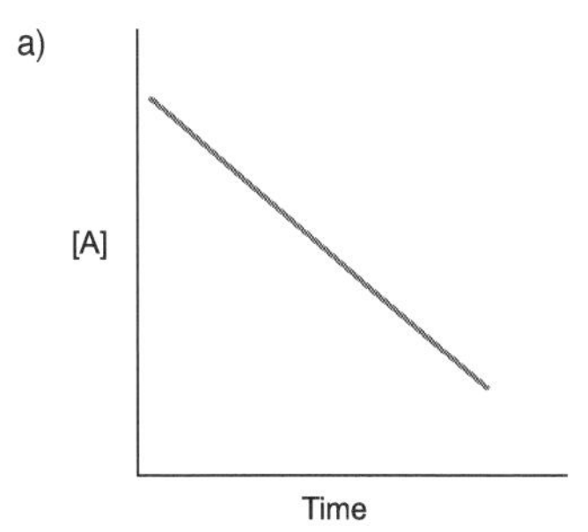

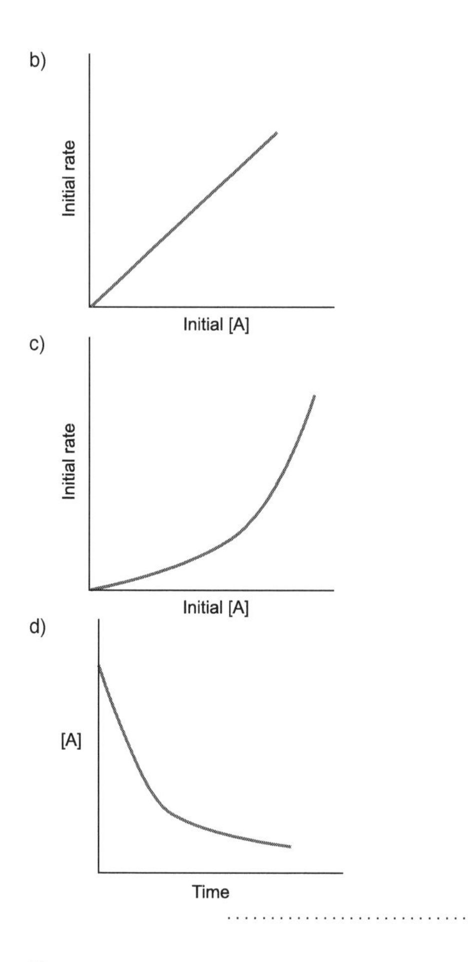

. . . .

Mercury(II) chloride is reduced by oxalate ions ($C_2O_4^{2-}$) according to the equation:

$2HgCl_2$ (aq) + $C_2O_4^{2-}$ (aq) $\rightarrow$ $2Cl^-$ (aq) + $2CO_2$ (g) + 2HgCl (s)

The following data were obtained in a series of four experiments at the same temperature.

The rate is measured by the decrease in concentration of $HgCl_2$ (aq) per minute.

Experiment	Initial concentration of $HgCl_2(aq)$ / mol l^{-1}	Initial concentration of $C_2O_4^{2-}$ (aq) / mol l^{-1}	Initial rate / mol l^{-1} min^{-1}
1	0.128	0.304	1.82×10^{-4}
2	0.064	0.608	3.66×10^{-4}
3	0.128	0.608	7.31×10^{-4}
4	0.064	0.304	0.90×10^{-4}

Q38: From the data given above, deduce the overall rate equation for the reaction.

....

Q39: Using the results for experiment 1 and your answer to part (a), calculate the rate constant at the given temperature.

....

Q40: What will be the units of k?

a) l mol^{-1} min^{-1}
b) mol min^{-1}
c) mol^2 l^{-2} min^{-1}
d) l^2 mol^{-2} min^{-1}
e) mol l^{-1} min^{-1}
f) l min^{-1}

....

Q41. Calculate the initial rate of the reaction when the initial concentration of each reactant is 0.1 mol l^{-1}.

....

....

Topic 8

End of Unit 1 test

Go online

End of Unit 1 test

Q1: Compared to visible radiation (visible light), infra-red radiation has a:

a) lower frequency.
b) higher velocity.
c) lower velocity.
d) higher frequency.

....

Q2: In an emission spectrum the frequency of each line corresponds to:

a) the energy change when an electron moves to a higher energy level.
b) the energy change when an electron moves to a lower energy level.
c) the kinetic energy possessed by an electron in an atom.
d) an energy level within an atom.

....

Q3: The emission spectrum of an element is seen as a series of bright coloured lines against a dark background. The brightest line in the emission spectrum of sodium is seen at 589 nm.

What causes a line in an emission spectrum?

....

Q4: Calculate the frequency of the emission line at 589 nm.

....

Q5: What is the electronic configuration of a vanadium atom?

a) $1s^22s^22p^63s^23p^63d^34s^2$
b) $1s^22s^22p^63s^23p^63d^44s^1$
c) $1s^22s^22p^63s^23p^63d^5$
d) $1s^22s^22p^63s^23p^64s^24p^3$

....

Q6: When electrons occupy degenerate orbitals, they do so in such a way as to maximise the number of parallel spins.

What is this statement known as?

a) The Pauli exclusion principle
b) The Aufbau principle
c) Hund's rule
d) Heisenberg's uncertainty principle

....

The 3d and 4s electrons for the iron atom can be represented as follows:

3d

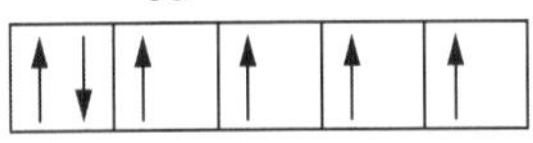

4s

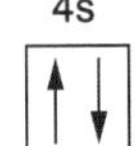

Q7: Complete a corresponding diagram for the Fe^{3+} ion.

	3d						4s
Fe^{3+}							

....

Q8: Suggest why the Fe^{3+} ion is more stable than the Fe^{2+} ion.

....

Q9: What is the shape of the NH_3 molecule?

a) Trigonal pyramidal
b) Trigonal planar
c) Square planar
d) Tetrahedral

....

Q10: NH_4^+ ion contains four bonding pairs of electrons.

What is the shape of the NH_4^+ ion?

a) Tetrahedral
b) Trigonal planar
c) Trigonal bipyramidal
d) Linear

....

Calculate the number of bonding and non-bonding pairs of electrons around the central atom in Cl_2O and hence work out the shape of the molecule.

Q11: How many bonding pairs are there?

....

Q12: How many non-bonding pairs are there?

....

Q13: The shape of the molecule Cl_2O is

a) angular
b) linear
c) octahedral
d) trigonal pyramidal
e) square planar
f) tetrahedral
g) trigonal planar
h) trigonal bipyramidal

....

Calculate the number of bonding and non-bonding pairs of electrons around the central atom in PCl_3 and hence work out the shape of the molecule.

Q14: How many bonding pairs are there?

....

Q15: How many non-bonding pairs are there?

....

Q16: The shape of the molecule PCl_3 is

a) angular
b) linear
c) octahedral
d) trigonal pyramidal
e) square planar
f) tetrahedral
g) trigonal
h) trigonal planar
i) trigonal bipyramidal

....

Calculate the number of bonding and non-bonding pairs of electrons around the central atom in SiF_4 and hence work out the shape of the molecule.

Q17: How many bonding pairs are there?

....

Q18: How many non-bonding pairs are there?

....

Q19: The shape of the molecule SiF_4 is

a) angular
b) linear
c) octahedral
d) trigonal pyramidal
e) square planar
f) tetrahedral
g) trigonal
h) trigonal planar
i) trigonal bipyramidal

....

Q20: What is the formula for the hexaamminetitanium(III) ion?

a) $[Ti(CH_3NH_2)_6]^{3-}$
b) $[Ti(NH_3)_6]^{3+}$
c) $[Ti(CH_3NH_2)_6]^{3+}$
d) $[Ti(NH_3)_6]^{3-}$

....

Q21: Name the $[CuCl_4]^{2-}$ ion.

....

Q22: What is the oxidation number of the iron in the complex ion $[Fe(CN)_6]^{3-}$?

a) -3
b) +2
c) +3
d) +6

....

Q23: $2SO_2 (g) + O_2 (g) \rightleftharpoons 2SO_3 (g)$

Removing the sulfur trioxide produced in the above system will:

a) increase the value of the equilibrium constant.
b) decrease the concentration of SO_2 and O_2.
c) increase the concentration of O_2 only.
d) decrease the value of the equilibrium constant.
e) decrease the concentration of SO_2 only.
f) increase the concentration of SO_2 and O_2.

....

Q24: Which of the following would **not** act as a buffer solution?

a) Ethanoic acid and sodium ethanoate
b) Hydrochloric acid and sodium chloride
c) Sulfurous acid and potassium sulfite
d) Aqueous ammonia and ammonium chloride

....

500 cm^3 of a buffer solution contains 0.20 mol of ethanoic acid and 0.25 mol of sodium ethanoate.

Q25: Explain how this solution acts as a buffer on the addition of a small volume of potassium hydroxide solution.

....

Q26: Using the above data and information from the data booklet, calculate the pH of the buffer solution using the equation pH = pKa - log ([acid] / [salt]).

....

Magnesium nitrate decomposes on heating.

$2Mg(NO_3)_2$ (s) $\rightarrow$ $2MgO$ (s) + $4NO_2$ (g) + O_2 (g) ΔH° = + 510 kJ mol^{-1}

Compound	**S° J K^{-1} mol^{-1}**
$Mg(NO_3)_2$ (s)	164
MgO (s)	27
NO_2 (g)	240
O_2 (g)	205

Q27: Calculate the standard entropy change for the above reaction.

ΔS°=

....

Q28: Calculate the temperature at which this reaction becomes feasible.

T=

....

Q29: At which temperature would the entropy of a perfect crystal be zero?

a) 298 K
b) 273 K
c) 100 K
d) 0 K

....

Q30: ΔG° gives an indication of the position of the equilibrium for a reaction.

The equilibrium lies on the side of the products when ΔG° is:

a) negative.
b) zero.
c) one.
d) positive.

....

Q31: The following reaction is first order with respect to each of the reactants.

A + B $\rightarrow$ C + D

Which of the following is a correct statement about this reaction?

a) As the reaction proceeds its rate will increase.
b) The rate of reaction is independent of the concentration of either A or B.
c) The reaction is second order overall.
d) The reaction is first order overall.

....

The following kinetic data was obtained for the reaction:
P + 3Q → R

[P] (mol l^{-1})	**[Q] (mol l^{-1})**	**Initial rate of formation of R (mol l^{-1} min^{-1})**
1.0	1.0	0.150
2.0	1.0	0.300
2.0	0.5	0.075

Q32: Deduce the rate equation for the reaction.

Rate = k

. . . .

Q33: Predict an equation for the rate determining step.

. . . .

Q34: Calculate the rate constant.

k=

. . . .

Q35: What will be the units of k?

a) l mol^{-1} min^{-1}
b) mol^2 l^{-2} min^{-1}
c) mol l^{-1} min^{-1}
d) l^2 mol^{-2} min^{-1}
e) l min^{-1}
f) mol min^{-1}

. . . .

Q36: The reaction A + B → C has a rate law of the form rate = k[A][B].

If the concentration of A and B are both doubled, the rate will increase by a factor of:

a) 2
b) 4
c) 6
d) 8

. . . .

. . . .

Glossary

Amphoteric

a substance which can act as both an acid and a base

Aufbau principle

this states that orbitals are filled in order of increasing energy

Avogadro's constant

is the number of constituent particles, usually atoms or molecules, that are contained in the amount of substance given by one mole ($L = 6.02 \times 10^{23}$ mol^{-1})

Buffer solution

a solution in which the pH remains approximately constant when small amounts of acid or base are added

Closed

a closed system has no exchange of matter or energy with its surroundings

Conjugate acid

the species left when a base accepts a proton

Conjugate base

the species formed when an acid donates a proton

Coordination number

the number of bonds a transition metal ion forms with surrounding ligands

Dative

a bond where both electrons have come from one of the elements involved in the bond

Degenerate

a set of atomic orbitals that are of equal energy to each other are said to be degenerate

Dynamic equilibrium

a dynamic equilibrium is achieved when the rates of two opposing processes become equal, so that no net change results

Electromagnetic spectrum

is the range of frequencies or wavelengths of electromagnetic radiation.

Enthalpy of formation

the enthalpy change when one mole of a substance is formed from its elements in their standard states

Entropy

the degree of disorder of a system

Equivalence point

the equivalence point in a titration experiment is reached when the reaction between the titrant (added from the burette) and the titrate (in the flask) is just complete.

Free energy

the total amount of energy available to do work

Frequency

is the number of wavelengths that pass a fixed point in one unit of time, usually one second.

Ground state

this is the lowest possible electronic configuration the electrons in an atom can adopt

Heisenberg's uncertainty principle

this states that it is impossible to state precisely the position and the momentum of an electron at the same instant

Hund's rule

when degenerate orbitals are available, electrons fill each singly, keeping their spins parallel before pairing starts

Ionic product of water

$Kw = [H^+][OH^-] = 1 \times 10^{-14}$ at 298 K

Ionisation energy

the first ionisation energy of an element is the energy required to remove one electron from each of one mole of atoms in the gas phase to form one mole of the positively charged ions in the gas phase

Ligand

an ion or molecule which can bind to a transition metal ion to form a complex: ligands have a negative charge or at least one lone pair of electrons

Pauli exclusion principle

this states that an orbital holds a maximum of two electrons with opposite spins (i.e. no two electrons can have the same set of four quantum numbers)

Planck's constant

is the physical constant that is the quantum of action in quantum mechanics ($h = 6.63 \times 10^{-34}$ J s)

Rate determining step

the slowest step in a reaction mechanism that governs the overall rate

Second law of thermodynamics

the total entropy of a reaction system and its surroundings always increases for a spontaneous change

Spectrochemical series

a list of ligands according to how strongly they split d orbitals in a transition metal complex. From largest to smallest splitting ability $CN^- > NO_2^- > NH_3 > H_2O > OH^- > F^- > Cl^- > Br^- > I^-$

Standard conditions

298 K (25 ^{0}C) and one atmosphere pressure

Velocity

is the physical vector quantity which needs both magnitude and direction to define it. Usually measured in m s^{-1} (or m/s)

Wavelength

is the distance between adjacent crests (or troughs) and is usually measured in metres or nanometres (1 nm = 10^{-9} m)

Answers to questions and activities

1 Electromagnetic radiation and atomic spectra

Answers from page 5.

Q1:

$$f = 2.45 \times 10^{9} \text{ Hz}$$
$$and\ \lambda = \frac{c}{f} = \frac{3 \times 10^{8} \text{ m s}^{-1}}{2.45 \times 10^{9} \text{ s}^{-1}}$$
$$\lambda = \frac{3 \times 10^{8}}{2.45 \times 10^{9}} \text{ m}$$
$$\lambda = 1.224 \times 10^{-1} \text{ m}$$
$$\lambda = 12.24 \text{ cm}$$

Answers from page 5.

Q2: 589

Wavelength from frequency (page 5)

Q3: 3.75×10^{-5} m

Q4: 4.29×10^{-7} m

Q5: 3.75×10^{-2} m

Answers from page 5.

Q6:

$$c = \text{ speed of light } = 3 \times 10^{8} \text{ m s}^{-1}$$
$$\lambda = \text{ wavelength} = 405 \times 10^{-9} \text{ m}$$
$$c = \lambda \times f$$
$$f = \frac{c}{\lambda}$$
$$f = \frac{3 \times 10^{8} \text{ m s}^{-1}}{405 \times 10^{-9} \text{ m}}$$
$$f = 7.41 \times 10^{14} \text{ s}^{-1}$$

Answers from page 6.

Q7: c) 9.23×10^{14}

Frequency from wavelength (page 6)

Q8: 1.58×10^{14} Hz

Q9: 1.30×10^{14} Hz

Q10: 2.00×10^{14} Hz

Electromagnetic radiation table (page 6)

Q11:

Quantity	Symbol	Unit	Description
velocity	c	ms^{-1}	rate of travel
wavelength	λ	m	wavecrest separation
frequency	f	Hz	wave cycles per second

Using spectra to identify samples (page 10)

Q12: Hydrogen

Q13: Helium

Q14: Calcium, since 650 nm has a triplet.

Q15: Sodium, as evidenced by the doublet at 580 nm.

Q16: Thallium

Q17: No, since there is no triplet at 580 nm.

Q18: No, since there is no triplet at 580 nm.

Q19: Thallium, since the line in sample B occurs at 530 nm and can only be credited to thallium from this database.

Answers from page 14.

Q20:

1. The wavelength of light needed:

Since $E = \frac{Lhc}{\lambda}$

and $E = 338$ kJ mol^{-1}

$$E = \frac{6.02 \times 10^{23} \text{mol}^{-1} \times 6.63 \times 10^{-34} \text{ J s} \times 3 \times 10^{8} \text{ m s}^{-1}}{\lambda}$$

$$\lambda = \frac{6.02 \times 10^{23} \text{mol}^{-1} \times 6.63 \times 10^{-34} \text{ J s} \times 3 \times 10^{8} \text{m s}^{-1}}{338 \times 10^{3} \text{ J mol}^{-1}}$$

$$\lambda = 0.3543 \times 10^{-6} \text{ m}$$

$$\lambda = 354.3 \text{ nm}$$

Major error is omitting multiplication by Avogadro's number.

Remember there are one mole of bonds.

2. These molecules can be unstable because this wavelength is within the ultraviolet region and sunlight, particularly in the upper atmosphere can provide this wavelength.

Energy from wavelength (page 14)

Q21: 74.8 kJ mol^{-1}

Q22: 92.1 kJ mol^{-1}

Q23: 85.5 kJ mol^{-1}

Q24: 63.0 kJ mol^{-1}

End of Topic 1 test (page 17)

Q25: a) γ - radiation

Q26: d) photons.

Q27: b) Colour moves towards red.

Q28: c) higher energy.

Q29: c) c

Q30: c) c*and* e) h, are both used to represent a constant

Q31: a) f

Q32: d) D

Q33: 408

Q34: Green

Q35: 223.8

2 Atomic orbitals, electronic configurations and the Periodic Table

Answers from page 25.

Q1:

Relating quantum numbers (page 28)

Q2:

Value of n	Value of l	Value of m	Subshell name
1	0	0	1s
2	0 1	0 -1 0 +1	2s 2p
3	0 1 2	0 -1 0 +1 -2 -1 0 +1 +2	3s 3p 3d

Spectroscopic notation (page 32)

Q3: c) $1s^2\ 2s^1$

Q4: d) Potassium

Q5: 4

Q6: 3

Q7: 1

Q8: d) Magnesium

Q9: b) $1s^2$

Orbital box notation (page 35)

Q10:

Electron configuration

Hydrogen (H) [↑] 1s

Helium (He) [↑↓] 1s

Lithium (Li)  [↑↓] 1s [↑] 2s

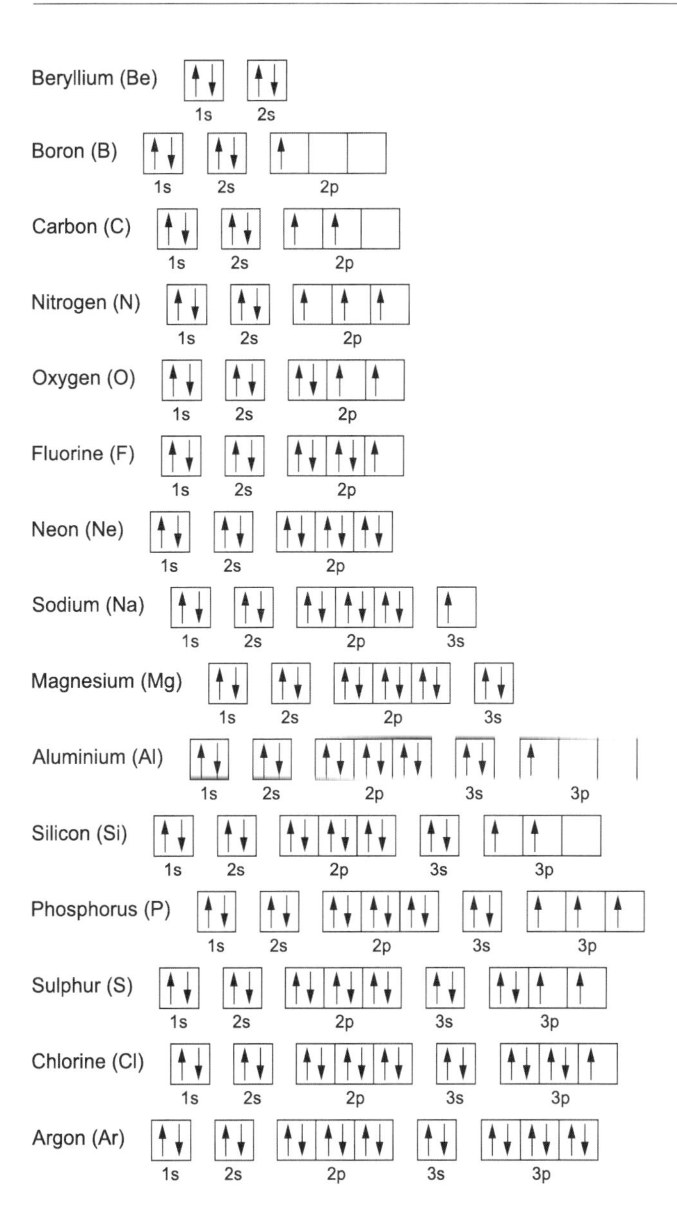
Beryllium (Be)
1s 2s
Boron (B)
1s 2s 2p
Carbon (C)
1s 2s 2p
Nitrogen (N)
1s 2s 2p
Oxygen (O)
1s 2s 2p
Fluorine (F)
1s 2s 2p
Neon (Ne)
1s 2s 2p
Sodium (Na)
1s 2s 2p 3s
Magnesium (Mg)
1s 2s 2p 3s
Aluminium (Al)
1s 2s 2p 3s 3p
Silicon (Si)
1s 2s 2p 3s 3p
Phosphorus (P)
1s 2s 2p 3s 3p
Sulphur (S)
1s 2s 2p 3s 3p
Chlorine (Cl)
1s 2s 2p 3s 3p
Argon (Ar)
1s 2s 2p 3s 3p

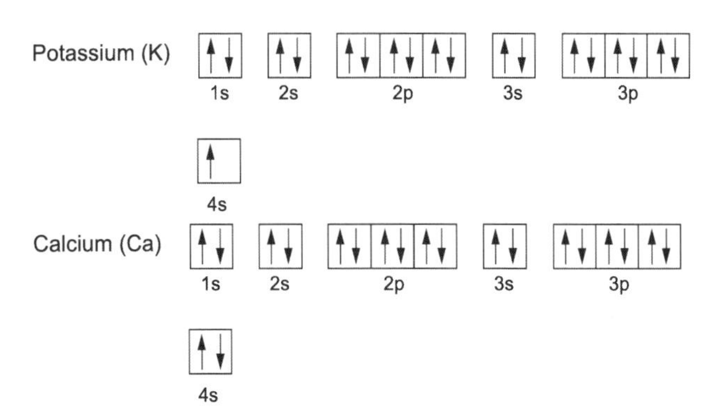

Periodic table blocks (page 36)

Q11:

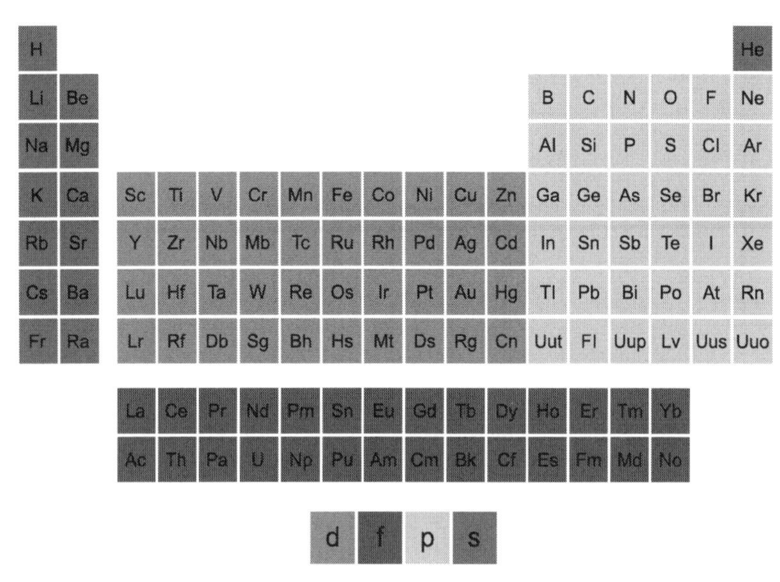

Q12: b) *p*

Q13: a) *s*

Q14: Transition

Q15: p

Q16: p

Q17: d

Q18: s

Ionisation energy evidence (page 39)

Q19: One

Q20: Easy

Q21: Reactive

Q22: One

First and second ionisation energies (page 40)

Q23: Sodium has **one electron** in its outer shell whereas magnesium has **two electrons** in its outer shell. The first ionisation energy of magnesium is **higher** than that of sodium since magnesium has 12 protons in its nucleus and therefore has a higher nuclear charge and a **stronger** attraction for the outer electrons. However, the second ionisation energy of **sodium** is higher than that of **magnesium** since the electrons being removed come from a **complete** p subshell which is **closer** to the nucleus.

End of Topic 2 test (page 42)

Q24: a) the Pauli exclusion principle.

Q25: d) Chlorine

Q26: d) 4

Q27: c) 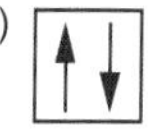 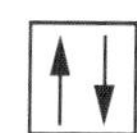 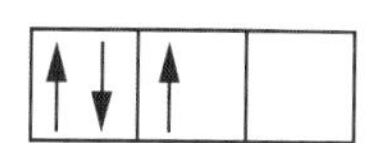

Q28: c) 3

Q29: b) would be needed to remove 1 mole of 2p electrons and 1 mole of 2s electrons.

Q30: a) $Mg^{+}(g) \rightarrow Mg^{2+}(g) + e^{-}$

Q31: 3 electrons

Q32: 1 electron

Q33: 3 quantum numbers

Q34: Degenerate

Q35: Argon

Q36: As the principal quantum number (n) increases, the energy levels become closer and closer together.

Q37: A phosphorus atom has three electrons in the 3p sub-shell, i.e. a half filled subshell which is relatively stable. A sulfur atom has four electrons in the 3p subshell, two unpaired and two paired. It is easier to remove one of the paired electrons from the sulfur atom than it is to remove an electron from the relatively stable phosphorus atom.

3 Shapes of molecules and polyatomic ions

Answers from page 48.

Q1: D

Q2: Bond length

Q3: Exothermic

Q4: The covalent radius for hydrogen is given as 37 pm and this is exactly half the bond length of the H-H bond since there are two hydrogens within it.

Q5: 136 pm since the covalent radius of chlorine is given as 99 pm and 99 + 37 = 136pm.

Resonance structures (page 50)

Q6: d) O_2

Q7: b) FCl

Q8: Methane and carbon dioxide have the following Lewis electron dot structures:

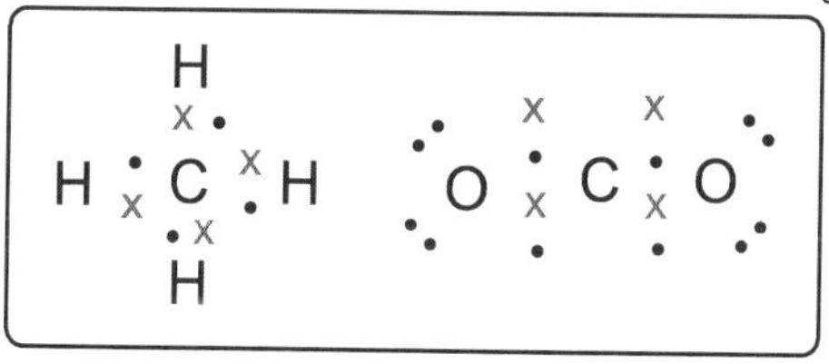

```
      H
      x •
H •x C x• H      •• O •x C x• O ••
      • x
      H
```

Q9: Carbon monoxide (showing shared pairs as straight lines) is:

$:C \equiv O:$

Summary of shapes of covalent molecules (page 55)

Q10:

Number of Electron Pairs	Arrangement	Angle(s) in degrees	Example
2	linear	180	$BeCl_2$
3	trigonal planar	120	BF_3
4	tetrahedral	109.5	CH_4
5	trigonal bipyramidal	90, 120, 180	PCl_5
6	octahedral	90	SF_6

Q11: a) PF_3

Q12: c) Trigonal bipyramid

Q13: d) BeF_2

Q14:

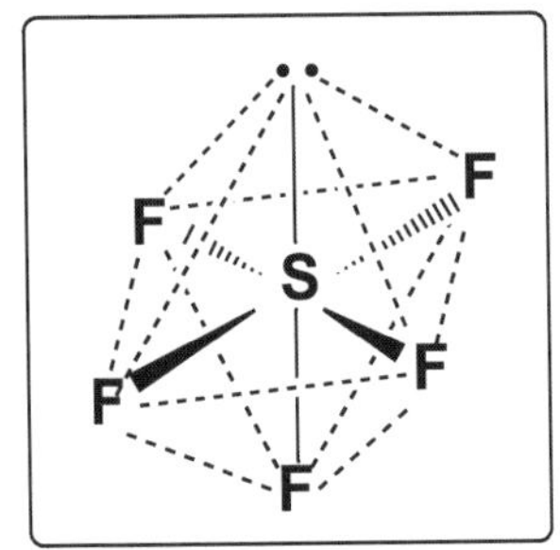

Q15: Octahedral

Q16: This shape is a square pyramid.

End of Topic 3 test (page 58)

Q17: d) Calcium oxide

Q18: d) phosphorus donating both electrons of the bond to boron.

Q19: c) H_2S

Q20: b) Trigonal planar to tetrahedral

Q21: b) 107°

Q22: b) Angular

Q23: d) Trigonal pyramidal

Q24: d) Trigonal pyramidal

Q25: a) Tetrahedral

Q26: c) Three negative

Q27: a) A

Q28:

F F B F $F^{\ominus}$ → $[F F B F F]^{\ominus}$

Dative covalent bond

Q29: b)

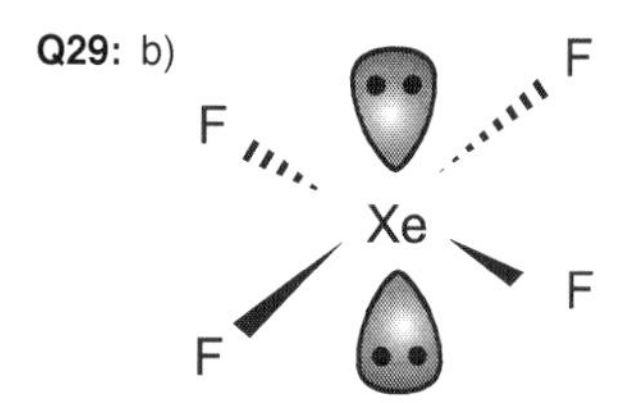

Q30: The arrangement of the bonding pairs is square planar.

4 Transition metals

Orbital box notation (page 64)

Q1:

Elements	Z	Electron configuration	Electron in box diagram
Scandium	21	$1s^2\ 2s^2\ 2p^6\ 3s^2\ 3p^6\ 4s^2\ 3d^1$	4s [↑↓] 3d [↑][ ][ ][ ][ ]
Titanium	22	$1s^2\ 2s^2\ 2p^6\ 3s^2\ 3p^6\ 4s^2\ 3d^2$	4s [↑↓] 3d [↑][↑][ ][ ][ ]
Vanadium	23	$1s^2\ 2s^2\ 2p^6\ 3s^2\ 3p^6\ 4s^2\ 3d^3$	4s [↑↓] 3d [↑][↑][↑][ ][ ]
Chromium	24	$1s^2\ 2s^2\ 2p^6\ 3s^2\ 3p^6\ 4s^1\ 3d^5$	4s [↑] 3d [↑][↑][↑][↑][↑]
Manganese	25	$1s^2\ 2s^2\ 2p^6\ 3s^2\ 3p^6\ 4s^2\ 3d^5$	4s [↑↓] 3d [↑][↑][↑][↑][↑]
Iron	26	$1s^2\ 2s^2\ 2p^6\ 3s^2\ 3p^6\ 4s^2\ 3d^6$	4s [↑↓] 3d [↑↓][↑][↑][↑][↑]
Cobalt	27	$1s^2\ 2s^2\ 2p^6\ 3s^2\ 3p^6\ 4s^2\ 3d^7$	4s [↑↓] 3d [↑↓][↑↓][↑][↑][↑]
Nickel	28	$1s^2\ 2s^2\ 2p^6\ 3s^2\ 3p^6\ 4s^2\ 3d^8$	4s [↑↓] 3d [↑↓][↑↓][↑↓][↑][↑]
Copper	29	$1s^2\ 2s^2\ 2p^6\ 3s^2\ 3p^6\ 4s^1\ 3d^{10}$	4s [↑] 3d [↑↓][↑↓][↑↓][↑↓][↑↓]
Zinc	30	$1s^2\ 2s^2\ 2p^6\ 3s^2\ 3p^6\ 4s^2\ 3d^{10}$	4s [↑↓] 3d [↑↓][↑↓][↑↓][↑↓][↑↓]

Answers from page 65.

Q2: Scandium only forms 3+ ions and Zinc only forms 2+ ions. Neither of these result in an incomplete d subshell, therefore do not fit the definition of a transition metal.

Q3: Fe^{3+} ions would have a half-filled d subshell which is stable.

Answers from page 66.

Q4: +5

Q5: +6

Q6: +4

Q7: +3

Q8: +6

Q9: +2

Q10: +6

Q11: +6

Q12: +3

Naming transition metal complexes (page 71)

Q13: Hexaaquacobalt(II) chloride

Q14: Sodium tetrafluorochromate(III)

Q15: Potassium hexacyanoferrate(II)

Q16: Potassium trioxalatoferrate(III)

Q17: 4

Q18: 6

Q19: 6

Q20: Octahedral

Q21: b) $Na_2[PtCl_4]$

Q22: d) $[Cu(CN)_2(H_2O)_2]$

Q23: a) $[CrCl(H_2O)_5]Cl_2$

Q24: c) $[CoCl_2(NH_3)_4]Cl$

Colour of transition metal compounds (page 74)

Q25: If violet is absorbed, **yellow** is transmitted.
If **red** is absorbed, green is transmitted.
If orange is absorbed, **blue** is transmitted.
When all colours of light are present **white** light is produced.

Q26: Blue

Q27: Blue

Q28: Yellow

Explanation of colour in transition metal compounds (page 77)

Q29: 164.0

Q30: b) Blue

Q31: 210.1

Q32: c) Violet

Q33: 257.5

Q34: b) Yellow

Q35: c) $Cl^- < H_2O < NH_3$

End of Topic 4 test (page 83)

Q36: a) $1s^2\ 2s^2\ 2p^6\ 3s^2\ 3p^6\ 3d^3\ 4s^2$

Q37: d)

Q38: d) Cr^{3+}

Q39: b) +2

Q40: c) Yellow

Q41: b) a reduction with gain of one electron.

Q42: b) The concentration of the absorbing species can be calculated from the intensity of the absorption.

Q43: a) Cation and d) Octahedral

Q44: f) Monodentate

Q45: Hexaamminechromium(III) chloride

Q46: The oxidation state of cobalt in this complex is +3 .

Q47: Ammonia (NH3) causes the stronger ligand field splitting.

Q48: The reaction speeds up when the cobalt(II) chloride is added.

Q49: c) Cobalt exhibits various oxidation states of differing stability.

Q50: $[Co(NH_3)_6]^{3+}$ ions are yellow (red and green mixed) which means that they must absorb blue light.
$[CoF_6]^{3-}$ ions are blue which means that they must absorb yellow light (red and green mixed).
Blue light is of higher energy than yellow light. So, ammonia ligands produce a greater splitting of the d orbitals than fluoride ions.

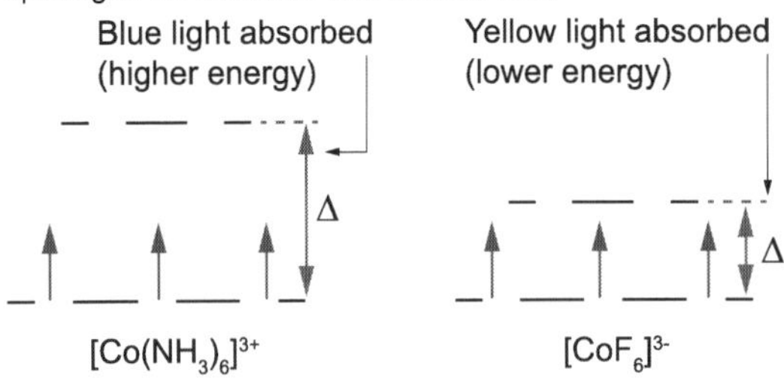

5 Chemical equilibrium

Hydrogen Iodide equilibrium (page 90)

Q1: At t = 15 there are 5, 5 and 2, respectively. At t = 30 and t = 70 there is no change with four of each.

Q2: There are four of each at both times.

Q3: They are the same.

Answers from page 93.

Q4:

$$Kc = \frac{\left[Fe^{2+}\right]^2 \left[I_3^-\right]}{[Fe^{3+}]^2[I^-]^3}$$

Q5:

$$Kc = \frac{[H^+]^2 \left[HPO_4^{2-}\right]}{[H_3PO_4]}$$

Answers from page 94.

Q6: c) Hydrogen iodide

Q7: a) Phosphorus(V) chloride

Q8: b) Mg

Q9: d) $2SO_2(g) + O_2(g) \rightleftharpoons 2SO_3(g)$ K_c at 636°C = 3343

Q10: The value increases from 21.1 to 3343 as the temperature drops.

Q11: a) 636°C

Answers from page 95.

Q12:

Q13:

$$Kp = \frac{(pSO_3)^2}{(pSO_2)^2\,(pO_2)}$$

Paper chromatography (page 99)

Q14: b) Black

Q15: e) Green

Q16: a) has the highest solvent/water partition coefficient.

Q17: c) Dark blue

Q18: They are probably the same dyes in both cases since the R_f values would be the same. If they were different materials, they would probably have moved different distances.

Calculating pH (page 105)

Q19: 2.30 pH

Q20: 5.10 pH

Q21: 12.80 pH

Q22: 10.46 pH

Q23:

$[H^+] = 5.00 \times 10^{-3}$ mol ℓ^{-1}

$[OH^-] = 2.00 \times 10^{-12}$ mol ℓ^{-1}

Q24:

$[H^+] = 2.51 \times 10^{-6}$ mol ℓ^{-1}

$[OH^-] = 3.98 \times 10^{-9}$ mol ℓ^{-1}

Q25:

$[H^+] = 3.98 \times 10^{-12}$ mol ℓ^{-1}

$[OH^-] = 2.51 \times 10^{-3}$ mol ℓ^{-1}

Q26:

$[H^+] = 1.26 \times 10^{-2}$ mol ℓ^{-1}

$[OH^-] = 7.94 \times 10^{-13}$ mol ℓ^{-1}

Answers from page 110.

Q27: 2.88

Answers from page 110.

Q28: The acid is nitric acid and the base is magnesium hydroxide.

Q29: The acid is hydrobromic acid and the base is potassium hydroxide.

Q30: The acid is ethanoic acid and the base is sodium hydroxide.

Q31: The acid is sulfurous acid (not sulfuric) and the base is calcium hydroxide.

Answers from page 111.

Q32: c) 9

Q33: Alkaline

Q34: d) The acid, hydrogen cyanide, is weak and the base is strong.

Q35: 9

Q36: If the pH is less than 7, then the acid is stronger than the base. So pyridine must be a weak base.

pH titration (page 114)

Q37: 7

Q38: At the equivalence point, the exact amount of alkali has been added to neutralise the acid; no more, no less.

Q39: 0.1

Q40: 3

Q41: d) The pH changes rapidly only around the equivalence point.

Titration curves (page 115)

Q42: a) Strong acid/strong alkali

Q43: c) 7

Q44: b) 5

Q45: d) 9

Q46: The pH at the equivalence point is the same as the pH of the salt formed.

combination	pH of salt
strong acid/strong alkali	7
strong acid/weak alkali	<7
weak acid/strong alkali	>7
weak acid/weak alkali	depends on relative strengths

Choosing indicators (page 117)

Q47: a) Suitable

Q48: a) Suitable

Q49: b) Unsuitable

Q50: b) Bromothymol blue

Q51: Bromothymol blue changes colour over the pH range 6.0-7.6 which contains the equivalence point. Phenolphthalein will also work well since the pH of the solution is changing rapidly over its pH range of around 8.0-10.0 For both these indicators adding a single drop of alkali at the end point should cause the colour change. Methyl orange is less suitable since a larger volume would be needed to cause its colour to change.

Q52: b) Unsuitable

Q53: a) Suitable

Q54: a) Suitable

Q55: d) Either methyl orange or bromothymol blue

Q56: For both methyl orange and bromothymol blue the pH of the solution is changing rapidly over the indicators' pH ranges. So there will be a sharp endpoint even although the equivalence point falls in neither range.

Q57: a) Suitable

Q58: b) Unsuitable

Q59: b) Unsuitable

Q60: a) Phenolphthalein

Q61: The pH of the equivalence point falls within the pH range over which phenolphthalein changes colour. So there will be a sharp endpoint. Both the other indicators will change colour gradually. For methyl orange, the colour change takes place long before the equivalence point.

Q62: b) Unsuitable

Q63: b) Unsuitable

Q64: b) Unsuitable

Q65: d) None of these

Q66: The pH change around the equivalence point is fairly gradual. In general, no indicator is suitable for the titration of a weak acid and a weak alkali. Such titrations have to be monitored using a pH meter.

Summary of buffer systems (page 125)

Q67:

Adding acid $H^+ + NH_3 \rightleftharpoons NH_4^+$

Adding alkali $OH^- + NH_4^+ \rightleftharpoons NH_3 + H_2O$

Q68: Adding acid H^+ + CH_3COO^- $\rightleftharpoons$ CH_3COOH

Adding alkali OH^- + CH_3COOH $\rightleftharpoons$ CH_3COO^- + H_2O

Buffer calculations (page 128)

Q69: 5.46

Q70: 1.78

Q71: 1.79×10^{-5}

Q72: 4.6

Q73: 4.51

Q74: 1.55×10^{-5}

Extra questions (page 128)

Q75: c) CH_3COO^-

Q76: d) HSO_4^-

Q77: d) H_2O

Q78: a) Greater

Q79: b) Decreases.

Q80: 7.00

Q81: 3.51

Q82: 7.40

Q83: a)

$$Kc = \frac{[H^+]\,[OH^-]}{[H_2O]}$$

Q84: c) $K_w = [H^+]\,[OH^-]$

Q85:

In **neutral water** the concentrations of the ionic species $[H^+]$ is equal to $[OH^-]$ and the value of K_w is 1.0×10 **-14** at 25°C.

In acidic solutions the concentrations of the ionic species $[H^+]$ is **greater than** $[OH^-]$ and the value of K_w is 1.0×10 **-14** at 25°C.

In alkaline solution the concentrations of the ionic species $[H^+]$ is **less than** $[OH^-]$ and the value of K_w is 1.0×10 **-14** at 25°C.

Q86: 3.3

Q87: 1.6×10^{-10}

Q88: DBCA

Q89: CADB

Q90: Acidic

Q91: Less

Q92: Greater

Q93: b) $H_2PO_4^-$

Q94: Phosphoric acid, H_3PO_4

Q95: b) Basic

Q96: Ammonia

Q97: Ammonium ion

Q98: a) Ammonia, NH_3,

Q99: b) Ammonium ion, NH_4^+,

Q100:

An **acidic** buffer solution contains a mixture of a **weak acid** and one of its **salts**. An example is a mixture of **ethanoic acid** and potassium ethanoate in water.

A **basic** buffer solution contains a mixture of a **weak base** and one of its **salts**. An example is a mixture of **ammonia** and ammonium chloride in water.

Q101:

Relative formula mass of ethanoic acid.

Formula **CH_3COOH**

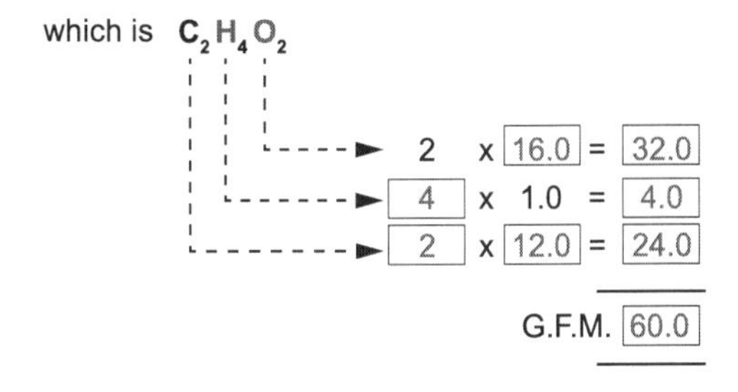

Q102: 6.0

Q103: 1.72×10^{-5}

Q104: 5.07

Q105: 4.47

Q106: 0.27

Q107: b) 100 cm^3 of 0.1 mol ℓ^{-1} HCOOH/0.2 mol ℓ^{-1} $HCOO^-Na^+$

Q108: c) Stay the same

End of Topic 5 test (page 141)

Q109: a) The reaction is endothermic.

Q110: b) Cu (s) + Mg^{2+} (aq) $\rightleftharpoons$ Cu^{2+} (aq) + Mg (s)

Q111: d) Increase of temperature

Q112: b) 8×10^{-2} atm

Q113: c) $$\frac{[NH_3]^2}{[H_2]^3[N_2]}$$

Q114: 1.8 mol l^{-1}

Q115: 1.4 mol l^{-1}

Q116: 0.032

Q117: The forward reaction is exothermic so when the temperature is raised the reaction will go in reverse to absorb heat (le Chatelier's principle). The value of K at this increased temperature will, therefore, be reduced.

Q118: d) mass of solute involved.

Q119: d) equal.

Q120: a) 0.5

Q121: 0.015 mol l^{-1}

Q122: 0.0075 mol l^{-1}

Q123: K = 2

Q124: c) C

Q125: c) C

Q126: b) Hexane

Q127: b) H_2O (l) + NH_3 (aq) $\rightarrow$ NH_4^+ (aq) + OH^- (aq)

Q128: d) NH_4^+ is the conjugate acid of NH_3.

Q129: c) 4.7

Q130: a) and d)

Q131: b) $K_w = [H^+] [OH^-]$

Q132: d) water will have a greater electrical conductivity at 25°C than at 18°C.

Q133: b) 0.50 mol l^{-1}

Q134: c) 10^{-10} and 10^{-11} mol l^{-1}

Q135: 3.4

Q136: 3.1

Q137: c) The overall colour of the solution depends on the ratio of [HIn] to [In⁻].

Q138: b) yellow in a solution of pH 3 and blue in a solution of pH 5.

Q139: c)

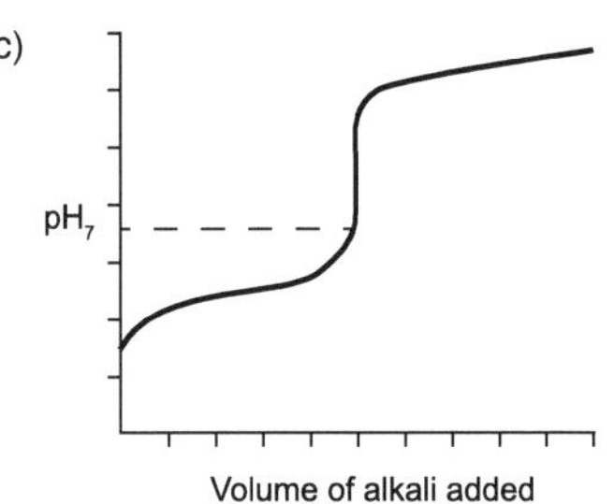

Q140: c) Phenolphthalein indicator, pH of colour change 8.0-9.8

Q141: a) The pH changes gradually around the equivalence point.

Q142: c) weak acid and a salt of that acid.

Q143: b) 50 cm^3 NH_4Cl (aq) + 50 cm^3 NH_3 (aq)

Q144: 3.47

Q145: 0.025 mol l^{-1}

Q146: c) Phenol red indicator, pH range 6.8-8.4

Q147: The salt sodium methanoate is formed in the reaction. Between E and F, some methanoic acid still remains unreacted and so there is a mixture of methanoic acid and sodium methanoate. A mixture of a weak acid and a salt of that acid is a buffer solution.

Q148: The methanoate ions present are able to remove added H+ ions. This creates methanoic acid.

6 Reaction feasibility

Answers from page 154.

Q1:

$\Delta S^o = \Sigma S^o_{(products)} - \Sigma S^o_{(reactants)}$
$= [(2 \times 43) + (2 \times 241) + 205] - (2 \times 142)$
$= 773 - 284$
$= 489\ J\ K^{-1}\ mol^{-1}$

The entropy values for Ag and O_2 are found on page 17 of the CFE Higher and Advanced Higher data book.

The entropy values for $AgNO_3$, NO_2 and Ag are multiplied by two due to the ratio of moles within the chemical equation.

Answers from page 155.

Q2: At 25°C (298 K), ΔS(total) is negative ($-436\ J\ K^{-1}\ mol^{-1}$)
At 1500°C (1773 K), ΔS(total) is positive ($+60.6\ J\ K^{-1}\ mol^{-1}$)

Q3: At 25°C (298 K), ΔS(total) = negative (approx $-10\ J\ K^{-1}\ mol^{-1}$)
At 5000°C (5273 K), ΔS(total) = negative (approx $-3.7\ J\ K^{-1}\ mol^{-1}$)
It is not thermodynamically feasible at either temperature.

Calculations involving free energy changes (page 158)

Q4: a) $2Mg(s) + CO_2(g) \rightarrow 2MgO(s) + C(s)$

$$\begin{aligned}\Delta G^o &= \Sigma G^o_{PRODUCTS} - \Sigma G^o_{REACTANTS} \\ &= (2 \times -569 + 0) - (2 \times 0 + (-394)) \\ &= -744\ kJ\ mol^{-1}\end{aligned}$$

The reaction where magnesium reduces carbon dioxide is feasible under standard conditions.

b) $2CuO(s) + C(s) \rightarrow 2Cu(s) + CO_2(g)$

$$\begin{aligned}\Delta G^o &= \Sigma G^o_{PRODUCTS} - \Sigma G^o_{REACTANTS} \\ &= (2 \times 0 + (-394)) - ((2 \times -130) + 0) \\ &= -134\ kJ\ mol^{-1}\end{aligned}$$

The reaction where carbon reduces copper(II) oxide is feasible under standard conditions.

Q5:

$$\Delta H^\circ = +117 \text{ kJ mol}^{-1}$$
$$\Delta S^\circ = +175 \text{ J K}^{-1} \text{ mol}^{-1}$$

Since $\Delta G^\circ = \Delta H^\circ - T\Delta S^\circ$

At 400 K $\Delta G^\circ_{400} = +47 \text{ kJ mol}^{-1}$

At 1000 K $\Delta G^\circ_{1000} = -58 \text{ kJ mol}^{-1}$

This reaction is feasible only at higher temperatures.

Q6:

$$\begin{aligned}
\Delta H^o &= -\ 92.8 \times 10^3\ J\ mol^{-1} \\
\Delta S^o &= -\ 198.6\ J\ K^{-1}\ mol^{-1} \\
when\ \Delta G^o &= 0 \\
T &= \frac{\Delta H^o}{\Delta S^o} \\
T &= 467.3\ K
\end{aligned}$$

Q7:

a) $\Delta G^\circ = +0.16$ kJ
b) Equilibrium position favours the reactants.

Q8: Boiling point = 333 K or 60°C

Interpreting Ellingham diagrams (page 163)

Q9: Silver(I) oxide. At 1000 K, ΔG is +60 kJ mol^{-1} on the graph. Even with no other chemical involved, this reverses to breakdown silver(I) oxide with ΔG° = -60 kJ mol^{-1}

Q10: Above 2200 K approximately. This would allow ΔG to be negative for:

$$2ZnO \rightarrow 2Zn + O_2$$

Q11: a)

(i) At 1000 K the target equation is:

$$2C(s) + 2ZnO(s) \rightarrow 2Zn(s) + 2CO(g)$$

(ii) $2C(s) + O_2(g) \rightarrow 2CO(g)$ $\Delta G^\circ = -400 \text{ kJ mol}^{-1}$

$2Zn(s) + O_2(g) \rightarrow 2ZnO(s)$ $\Delta G^\circ = -500 \text{ kJ mol}^{-1}$

(iii) Reverse the zinc equation:

$2ZnO(s) \rightarrow 2Zn(s) + O_2(g)$ $\Delta G^\circ = +500 \text{ kJ mol}^{-1}$

(iv) Adding to the carbon equation gives:

$2C(s) + O_2(g) + 2ZnO(s) \rightarrow 2CO(g) + 2Zn(s) + O_2(g)$

The oxygen on each side cancels out giving:

$2C(s) + 2ZnO(s) \rightarrow 2CO(g) + 2Zn(s)$

$\Delta G^\circ = +100$ kJ mol^{-1}

So at 1000 K the reaction is **not** feasible.

b)

(i) At 1500 K the target equation is the same:

$$2C(s) + 2ZnO(s) \rightarrow 2Zn(s) + 2CO(g)$$

(ii) $2C(s) + O_2(g) \rightarrow 2CO(g)$ $\Delta G^\circ = -500$ kJ mol^{-1}

$2Zn(s) + O_2(g) \rightarrow 2ZnO(s)$ $\Delta G^\circ = -300$ kJ mol^{-1}

(iii) Reverse the zinc equation:

$2ZnO(s) \rightarrow 2Zn(s) + O_2(g)$ $\Delta G^\circ = +300$ kJ mol^{-1}

(iv) Adding to the carbon equation gives:

$2C(s) + O_2(g) + 2ZnO(s) \rightarrow 2CO(g) + 2Zn(s) + O_2(g)$

The oxygen on each side cancels out giving:

$2C(s) + 2ZnO(s) \rightarrow 2CO(g) + 2Zn(s)$

$\Delta G^\circ = -200$ kJ mol^{-1}

So at 1500 K the reaction is feasible.

Q12: Where the two lines cross, $\Delta G^\circ = 0$. Above this temperature, the reaction is feasible. Approximately 1200K.
Remember: The lower of the two lines operates as written and the upper line will be reversed.

Q13: As the zinc melts, the disorder (entropy) increases. Since the gradient is given by $-\Delta S$ (from the straight line $\Delta G = -T\Delta S + \Delta H$), the slope of the line changes.

Q14: Zinc vaporises at 1180 K with an increase in entropy and a subsequent change in the gradient of the line on the Ellingham diagram.

Answers from page 164.

Q15:

a) $2FeO(s) + 2C(s) \rightarrow 2Fe(s) + 2CO(g)$
b) $\Delta G^\circ = -155$ kJ mol^{-1}
c) Above 1010 K
d) Below 980 K
e) It is a gas and can mix better with the solid iron(II) oxide.

Q16:

a) above about 2100 K
b) The high cost of maintaining temperature. The fact that magnesium is a gas at this temperature.
c) ΔG° = +160 kJ mol^{-1} (there would be some leeway in this figure).
d) ΔG° = +68 kJ mol^{-1} (dependent on your answer to part (c)).
e) Keeps the equilibrium reaction below from going in the reverse direction.

$$2MgO + Si \leftrightharpoons SiO_2 + 2Mg$$

End of Topic 6 test (page 168)

Q17: d) $N_2(g) + 3H_2(g) \rightarrow 2NH_3(g)$

Q18: c) -67.5 kJ mol^{-1} of ZnO

Q19: c) -242

Q20: c) large and negative.

Q21: c) 461 K

Q22: b) calcium.

Q23: The difference in the values of ΔG and ΔH is determined by the entropy change of the system.
The second reaction involves a reactant, gaseous oxygen, with a high entropy forming a solid, aluminium oxide, with a low entropy.

Q24: A reaction with a negative value for ΔG° will occur spontaneously, this has a negative ΔG° value.
A positive ΔH° indicates an endothermic reaction that will take heat from the surroundings which will drop in temperature.

Q25: 10.1 J K^{-1}mol^{-1}

Q26: +34.3

7 Kinetics

Orders and rate constants (page 173)

Q1: Rate = k $[Br^-][BrO_3^-][H^+]^2$

Q2: 4

Q3: 2

Q4: 0

Q5: 1

Q6: 1

Q7: Rate = k $[N_2O_5]$

Q8: s^{-1}

Q9: 0.00044 (normal decimal form)
4.4×10^{-4} (standard form)

Q10: 0.0000308 (normal decimal form)
3.08×10^{-5} (standard form)

Q11:

Rate = k $[H_2O_2][I^-]$

or

Rate = k $[H_2O_2][I^-][H^+]^0$

Q12: $mol^{-1}\ \ell\ s^{-1}$

Q13: 0.023

Answers from page 176.

Q14: 3001.5

Q15: 30001.5 seconds.

The filler takes 30 s to fill one bottle, so the thousandth bottle will be filled after 1000 $\times$ 30 s and a further 1.5 s will be needed to cap and label it.

Questions on reaction mechanisms (page 177)

Q16: b) $H_2S + Cl_2 \rightarrow S + 2HCl$

Q17: a) $4HBr + O_2 \rightarrow 2H_2O + 2Br_2$

Q18: b) False

Q19: b) False

Q20: a) True

Q21: b) False

Q22: b) False

Q23: a) True

Q24: $2H_2O_2 \rightarrow 2H_2O + O_2$

Q25: catalyst

Q26: intermediate

Q27: b) Rate = k $[H_2O_2][Br^-]$

End of Topic 7 test (page 181)

Q28: c) mol l^{-1} s^{-1}

Q29: a) Rate = k [X] [Y]

Q30: a) mol^{-1} l s^{-1}

Q31: d) the rate expression cannot be predicted.

Q32: d) 8

Q33: b) The rate of reaction decreases as the reaction proceeds.

Q34: c) **X + Y → intermediate**

Q35: d) $k[N_2O_5]$

Q36: c) **P + Q → R + S slow**
R + Q → T fast

Q37: a)

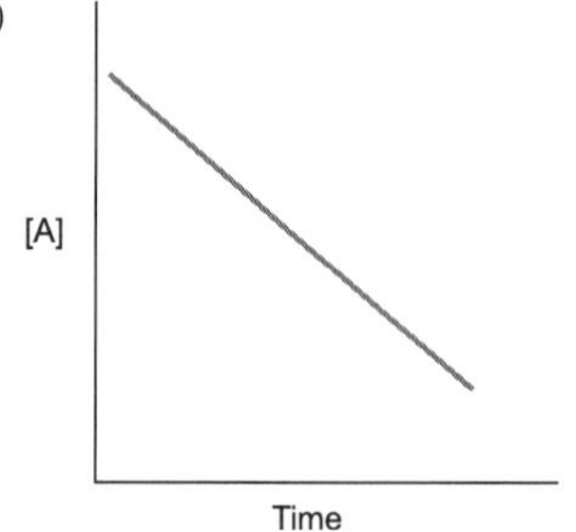

Q38: Rate = k $[C_2O_4{}^{2-}]2$ $[HgCl_2]$
One mark each for order with respect to each reactant.

Q39: 0.0154 mol^{-2} l^2 min^{-1}
1 mark for numerical answer, 1 mark for units

Q40: d) l^2 mol^{-2} min^{-1}

Q41: 1.54×10^{-5} mol l^{-1} min^{-1}

8 End of Unit 1 test

End of Unit 1 test (page 188)

Q1: a) lower frequency.

Q2: b) the energy change when an electron moves to a lower energy level.

Q3: Movement of electrons from higher to lower energy levels.

Q4: 5.09×10^{14}Hz

Q5: a) $1s^2 2s^2 2p^6 3s^2 3p^6 3d^3 4s^2$

Q6: c) Hund's rule

Q7:

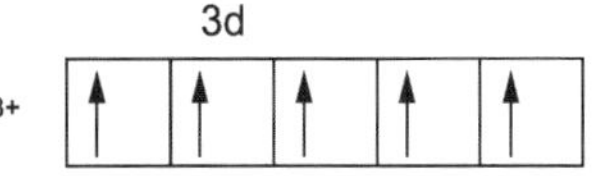

Q8: Stability of having all d-orbitals half-filled.

Q9: a) Trigonal pyramidal

Q10: a) Tetrahedral

Q11: 2

Q12: 2

Q13: a) angular

Q14: 3

Q15: 1

Q16: d) trigonal pyramidal

Q17: 4

Q18: 0

Q19: f) tetrahedral

Q20: b) $[Ti(NH_3)_6]^{3+}$

Q21: Tetrachlorocuprate (II)

Q22: c) +3

Q23: b) decrease the concentration of SO_2 and O_2.

Q24: b) Hydrochloric acid and sodium chloride

Q25: Added hydroxide ions react with the hydrogen ions and ethanoic acid molecules further dissociate to replace these hydrogen ions.

$H^+ + OH^- \leftrightharpoons H_2O$ (l)

$CH_3COOH \leftrightharpoons H^+ + CH_3COO^-$

Q26: 4.86

Q27: 891 J $K^{-1}mol^{-1}$

Q28: 572 K

Q29: d) 0 K

Q30: a) negative.

Q31: c) The reaction is second order overall.

Q32: $[P][Q]^2$

Q33: $P + 2Q \rightarrow$ Intermediate

Q34: 0.15

Q35: d) l^2 mol^{-2} min^{-1}

Q36: b) 4